Study Guide

Essentials of Meteorology

FIFTH EDITION

C. Donald Ahrens

Emeritus
Modesto Junior College

THOMSON
™
BROOKS/COLE

Australia • Brazil • Canada • Mexico • Singapore • Spain • United Kingdom • United States

Printed in the United States of America

1 2 3 4 5 6 7   11 10 09 08 07

Printer: Thomson/West

ISBN-13: 978-0-495-11902-9
ISBN-10: 0-495-11902-4

Thomson Higher Education
10 Davis Drive
Belmont, CA 94002-3098
USA

For more information about our products, contact us at:
Thomson Learning Academic Resource Center
1-800-423-0563

For permission to use material from this text or product, submit a request online at
http://www.thomsonrights.com.
Any additional questions about permissions can be submitted by email to **thomsonrights@thomson.com.**

Contents

Preface

This workbook/study guide has been developed as a tool to help you learn the information in the textbook, *Essentials of Meteorology*, 5th Edition. Meteorology is a complex subject; understanding weather and climate means mastering many new terms, concepts and ideas. It is my hope that with the aid of this workbook/study guide, learning about the weather around you will be a positive and enjoyable experience.

The workbook/study guide is organized so that each chapter begins with a brief *summary* of the information presented in the corresponding textbook chapter. This is followed by a list of the chapter's *important concepts and facts*. Continuing through the workbook/study guide you will find *matching, fill-in-the-blank, multiple choice, true-false* and *additional questions* that pertain to the chapter. I feel the best way to answer these questions is to first carefully read the chapter in your textbook, then begin answering the questions in the study guide. If you find that you are having difficulty with a certain concept, return to your textbook and reread the section dealing with that particular topic. When answering the *true-false* questions it is important that you know why an answer is false, not just that it is.

Toward the end of each chapter is a list of *additional readings*. They provide a closer look at certain issues and subjects associated with the topics covered in the chapter. Most are found in journals that are available in university and college libraries. I have also listed a few books that may be helpful.

The study of the science of meteorology can be fascinating and exciting. It is my hope that through your text, this study guide, and the information you gain in class you will come to a special appreciation of the dynamics of the atmosphere around you.

The Earth's Atmosphere

Chapter One provides you with a broad overview of the earth's atmosphere. The first part of this chapter begins with an examination of the various constituents found in today's atmosphere, including the different greenhouse gases and some of the major atmospheric pollutants. The next section describes one possible theory as to how the earth's atmosphere may have evolved. This is followed by a section that investigates the different layers of the atmosphere. Here we investigate the important concepts of pressure, density, and temperature, and how they vary with height. Then comes an overview of the earth's atmosphere that gives you a glimpse of several topics that will be covered more completely in later chapters, such as storm systems, weather maps, and satellite images. The last section briefly details the many ways weather and climate influence our lives, from how we feel on a windy day to the type of clothing we should purchase for the coming season.

Some important concepts and facts of this chapter:

1. In a given volume of dry air near the earth's surface, nitrogen occupies about 78 percent and oxygen about 21 percent.

2. Water is the only substance in our atmosphere that is found naturally as a liquid (water), as a gas (water vapor), and as a solid (ice).

3. Carbon dioxide (CO_2), an important greenhouse gas in the earth's atmosphere, has increased in concentration by more than 20 percent since 1958.

4. Other important greenhouse gases in the earth's atmosphere include water vapor (H_2O), methane (CH_4), nitrous oxide (N_2O), and chlorofluorocarbons (CFCs).

5. The majority of water on our planet is believed to have come from the earth's hot interior through outgassing.

6. Atmospheric pressure is a measure of the total mass of air above any point. Because of this fact, atmospheric pressure always decreases with increasing height above the ground.

7. The rate at which the air temperature decreases with height is called the *lapse rate*, whereas a measured *increase* in air temperature with height is called an *inversion*.

8. The majority of ozone (O_3) in our atmosphere is found in the stratosphere, the layer above the troposphere.

9. Weather is the condition of the atmosphere at any particular time and place. Climate is the accumulation of daily and seasonal weather events that occur over a given period of time.

10. The wind direction is the direction *from which* the wind is blowing. A north wind blows *from* the north.

11. The earth's atmosphere contains storms of all sizes, ranging from huge middle latitude cyclonic storms that can extend for thousands of kilometers, to much smaller tornadoes that are normally less than one kilometer wide.

Self Tests

Match the Following

_____ 1. Weather element that *always* decreases with increasing height

_____ 2. The most abundant greenhouse gas in the earth's atmosphere

_____ 3. Layer of the atmosphere that contains almost all the weather

_____ 4. The outpouring of gases from the earth's hot interior

_____ 5. Gas that strongly absorbs ultraviolet (UV) radiation in the stratosphere

_____ 6. The average decrease in air temperature with increasing height above the surface

_____ 7. A measured increase in air temperature with increasing height

_____ 8. The electrified region of the upper atmosphere

_____ 9. The study of the atmosphere and its phenomena

_____ 10. Another name for a high pressure area

_____ 11. A storm of tropical origin with winds in excess of 64 knots (74 mi/hr)

_____ 12. The horizontal movement of air

_____ 13. A relatively small, rotating funnel that extends downward from the base of a thunderstorm

_____ 14. It holds a planet's atmosphere close to its surface

_____ 15. On a weather map, this zone marks sharp changes in temperature, humidity and wind direction

a. outgassing

b. ionosphere

c. water vapor

d. anticyclone

e. lapse rate

f. wind

g. troposphere

h. gravity

i. inversion

j. tornado

k. meteorology

l. pressure

m. front

n. hurricane

o. ozone

Fill in the Blank

1. The gas that shows the most variation from place to place and from time to time in the lower atmosphere is _____ _____.

2. What percent does each of the following gases occupy in a volume of air near the earth's surface?

   nitrogen _____%

   oxygen _____%

   water vapor _____%

   carbon dioxide _____%

3. Most of the ozone in the atmosphere is found in the atmospheric layer called the _____.

4. The primary source of energy for the earth's atmosphere is the _____.

5. The hottest atmospheric layer is the _____.

6. The only substance near the earth's surface that is found naturally in the atmosphere as a solid, a liquid, and a gas is _____.

7. The atmospheric layer in which we live is called the _____.

8. The instrument that measures temperature, pressure, and humidity at levels above the earth's surface is the _____.

Multiple Choice

1. Which below is *not* a process by which CO_2 enters the atmosphere?

   a. volcanic eruptions
   b. decay of vegetation
   c. condensation
   d. burning of fossil fuels

2. The *primary* source of oxygen for the earth's atmosphere during the past half billion years or so appears to be:

   a. volcanic eruptions
   b. exhalation of animal life
   c. photosynthesis

3. Air density normally:

   a. increases with increasing height
   b. decreases with increasing height
   c. remains constant with increasing height

4. If you are standing north of a smoke stack and smoke from the stack is drifting over your head, the wind would be called a _____ wind.

    a. north
    b. south

5. The largest storm in our atmosphere in terms of actual size (diameter) is the:

    a. hurricane
    b. tornado
    c. middle latitude cyclonic storm
    d. thunderstorm

6. In the Northern Hemisphere, surface winds tend to blow this way around an area of *surface low pressure*.

    a. clockwise and inward
    b. clockwise and outward
    c. counterclockwise and inward
    d. counterclockwise and outward

7. Which of the following statements relates to weather rather than climate?

    a. The winters here are cold and wet.
    b. Outside it is sunny and hot.
    c. Thunderstorms are prevalent during July.
    d. Our lowest temperature ever, was –30°C.
    e. The average temperature during March is 20°C.

8. Of these four storms, the smallest in terms of actual size (diameter) is the:

    a. hurricane
    b. middle latitude cyclonic storm
    c. thunderstorm
    d. tornado

True–False

_____ 1. Over the last 100 years, the concentration of CO_2 in the earth's atmosphere has been increasing.

_____ 2. Water vapor is a gas.

_____ 3. The average air temperature at the tropopause is warmer than the average air temperature measured at the earth's surface.

_____ 4. The two most abundant gases in the stratosphere are nitrogen and oxygen.

_____ 5. About half of all the molecules in our atmosphere are below an altitude of about 5.5 kilometers or 18,000 feet.

_____ 6. A south wind blows from the south.

_____ 7. Generally, weather in the middle latitudes moves from east to west.

_____ 8. In the Northern Hemisphere, surface winds tend to blow clockwise and outward around an area of high pressure.

_____ 9. In our atmosphere, tiny solid or liquid suspended particles are called aerosols.

_____ 10. The atmosphere is compressible. This fact means that air pressure decreases at a constant rate from the Earth's surface to the top of the atmosphere.

Additional Questions

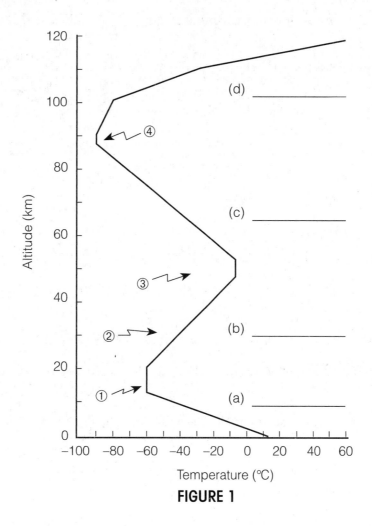

FIGURE 1

1. The following questions refer to Figure 1.

   a. In the diagram label the four layers of the atmosphere in the blank spaces provided.

   b. The atmospheric pressure would be lowest at number _____.

   c. Which number is pointing to the tropopause?
         a. 1   b. 2   c. 3   d. 4

   d. Which number is pointing to a temperature inversion?
         a. 1   b. 2   c. 3   d. 4

2.   In the adjacent weather map:

    a.   A warm front is positioned between numbers
        _____ and _____.

    b.   A cold front is positioned between numbers
        _____ and _____.

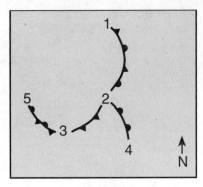

FIGURE 2

Additional Readings

"From Gods to Satellites" by John Farrand, Jr., *Weatherwise*, Vol. 44, No. 2 (April 1991), p. 30.

"To Understand the Atmosphere" by Henry Lansford, *Weatherwise*, Vol. 38, No. 4 (August 1985), p. 184.

"An Age of Discovery," *Weatherwise*, Vol. 48, No. 3 (June/July 1995), p. 42.

"The Great Leap Forward" by Patrick Hughes, *Weatherwise*, Vol. 47, No. 5 (October/November 1994), p. 22.

"Building BLOCKS," Steven Horstmeyer, *Weatherwise*, Vol. 54, No. 5 (September/October 2001), p. 20.

"Afraid of the Weather?" Ronald A. Kleinknecht and K. Bergren Smith, *Weatherwise*, Vol. 55, No. 6 (November/December 2002), p. 14.

"A Cool Early Earth," by John W. Valley, *Scientific American*, Vol. 293, No. 4 (October 2004), p. 58.

"On the Job," *Weatherwise*, Vol. 5, No. 6 (November/December 2004), p. 47.

"Making Records," by Randy Gerveny, *Weatherwise*, Vol. 59, No. 1 (January/February 2006), p. 48.

"The Case of the Missing Carbon," by Tim Appenzeler, *National Geographic*, Vol. 205, No. 2 (February 2004), p. 88.

"Seal of Approval," by Andrew Freedman, *Weatherwise*, Vol. 59, No. 1 (January/February 2006), p. 28.

Answers

Matching

1.	l	5.	o	9.	k	13.	j
2.	c	6.	e	10.	d	14.	h
3.	g	7.	i	11.	n	15.	m
4.	a	8.	b	12.	f		

Fill in the Blanks

1. water vapor
2. nitrogen = 78%
   oxygen = 21%
   water vapor = 0 to 4%
   carbon dioxide = 0.038%
3. stratosphere
4. sun
5. thermosphere
6. water
7. troposphere
8. radiosonde

Multiple Choice

1.	c	3.	b	5.	c	7.	b
2.	c	4.	b	6.	c	8.	d

True–False

1.	T	4.	T	7.	F	10.	F
2.	T	5.	T	8.	T		
3.	F	6.	T	9.	T		

Additional Questions

1. a. (a) troposphere
      (b) stratosphere
      (c) mesosphere
      (d) thermosphere
   b. number 4
   c. number 1
   d. number 2

2. a. A warm front is between number 2 and number 4
   b. A cold front is between number 2 and number 3

Warming the Earth and the Atmosphere

Chapter Two begins by examining the concepts of energy, temperature and heat. It then considers how heat energy is transferred in our atmosphere. Next, it covers the topics of absorption and emission of energy in terms of heating and cooling the earth. At this point we examine the atmospheric greenhouse effect and the various gases that produce it. Here we learn that without a greenhouse effect, the average surface temperature of our planet would be considerably colder than it is now. We also see that it is the enhancement of the greenhouse effect by the increasing concentrations of greenhouse gases that is of concern to most scientists. The section that follows describes how the earth and its atmosphere are warmed. Included here is a section on the energy balance of the earth and its atmosphere. Toward the end of the chapter is a discussion on seasonal temperature variations in the Northern and Southern Hemispheres. Here we learn that our seasons are caused by the earth being tilted on its axis as it revolves around the sun. The chapter concludes with a section that describes how seasonal variations in solar energy can influence temperatures on a much smaller scale, such as on the north and south side of a hill.

Some important concepts and facts of this chapter:

1. The temperature of a substance is a measure of the average speed of its molecules.

2. The transfer of heat within our atmosphere can take place by conduction, convection, and radiation.

3. Evaporation is a cooling process and condensation is a warming process.

4. Latent heat is an important source of atmospheric energy.

5. Rising air expands and cools, while sinking air is compressed and warms.

6. All objects with a temperature above absolute zero, –273°C (–459°F), emit radiation.

7. The higher an object's temperature, the more total radiation emitted each second by the object and the shorter are the wavelengths of emitted radiation.

8. The earth's surface absorbs solar radiation only during the daylight hours; however, it constantly emits infrared radiation, both during the day and at night.

9. Water vapor and carbon dioxide are important atmospheric greenhouse gases that selectively absorb and emit infrared radiation, thereby keeping the earth's average surface temperature warmer than it would be otherwise.

10. Enhancement of the atmospheric greenhouse effect is taking place because of the increasing concentrations of greenhouse gases.

11. The lower part of our atmosphere is mainly heated from the ground upward.

12. The annual average temperature of the earth and the atmosphere remains fairly constant from one year to the next because the amount of energy they absorb each year is equal to the amount of energy they lose.

13. The seasons are caused by the earth being tilted on its axis as it revolves around the sun. This tilt causes seasonal variations in both the length of daylight and the intensity of sunlight that reaches the surface.

14. When the Northern Hemisphere experiences winter (Dec., Jan., and Feb.), the Southern Hemisphere experiences summer (and vice versa).

15. South-facing sides of hills tend to be warmer and drier than their north-facing counterparts.

Self Tests

. . . .

Match the Following

_____ 1. Heat transfer process that depends upon the movement of air

_____ 2. Objects that selectively absorb and emit radiation

_____ 3. Rising bubbles of air

_____ 4. The heat we can feel and measure with a thermometer

_____ 5. The horizontal transfer of any atmospheric property by the wind

_____ 6. Energy transferred by electromagnetic waves

_____ 7. One millionth of a meter

_____ 8. A measure of the average speed of air molecules

_____ 9. The horizontal distance between two wave crests

_____ 10. This is released as sensible heat during the formation of clouds

_____ 11. The transfer of heat by molecule-to-molecule contact

_____ 12. The sun emits radiation with greatest intensity in this region of the spectrum

_____ 13. A temperature scale where 0° represents freezing and 100° boiling

_____ 14. Wavelengths longer than those of red light

_____ 15. Visible light given off by excited atoms and molecules in the upper atmosphere

_____ 16. Electromagnetic waves whose wavelengths are shorter than those of visible light

_____ 17. Temperature scale that begins at absolute zero

a. conduction

b. sensible heat

c. infrared

d. selective absorber

e. temperature

f. Kelvin

g. convection

h. aurora

i. radiation

j. wavelength

k. visible

l. thermals

m. latent heat

n. micrometer

o. ultraviolet

p. advection

q. Celsius

Additional Matching (The Seasons)

_____ 1. The astronomical beginning of fall

_____ 2. The day with the fewest hours of daylight in the Northern Hemisphere

_____ 3. The latitude at which days and nights are always of equal length

_____ 4. The day when, at noon, the sun is at its highest position in the Northern Hemisphere

_____ 5. The astronomical beginning of spring

_____ 6. An unseasonably warm spell with clear weather usually near the middle of autumn

a. winter solstice

b. vernal equinox

c. autumnal equinox

d. Indian summer

e. summer solstice

f. equator

Fill in the Blank

1. Energy of motion is also known as _____ _____.

2. Sunlight that bounces off a surface is said to be _____.

3. A perfect absorber and a perfect emitter of radiation is called a _____ .

4. How much radiation would an object be emitting if its temperature were at absolute zero? _____

5. The _____ represents the reflectivity of a surface.

6. The two most significant atmospheric greenhouse gases in the earth's atmosphere are _____ and _____.

7. At night objects on the ground cool by the process of emitting _____ _____.

8. The combined albedo of the earth and its atmosphere averages about _____ percent.

9. During January, it is winter in the Northern Hemisphere, and _____ in the Southern Hemisphere.

10. The sun emits maximum radiation in the _____ portion of the spectrum, while the earth emits maximum radiation at _____ wavelengths.

11. If the present concentration of CO_2 doubles, climatic models predict that for the earth's average temperature to increase by as much as 4.5°C, the gas _____ _____ must also increase in concentration.

12. The sun will be directly above Honolulu, Hawaii (latitude 21°N) _____ time(s) each year.

13. At the North Pole, the sun rises above the horizon on the vernal equinox and stays above the horizon until the _____ _____.

14. The wavelength range where neither water vapor nor carbon dioxide absorbs much of the earth's infrared radiation is known as the atmospheric _____.

15. Air that sinks, warms by _____.

Multiple Choice

1. As the average speed of air molecules decreases, the temperature of the air:

    a. increases
    b. decreases
    c. does not change

2.   The proper order of waves from longest to shortest is:

    a.   visible, infrared, ultraviolet
    b.   infrared, visible, ultraviolet
    c.   ultraviolet, visible, infrared
    d.   visible, ultraviolet, infrared
    e.   ultraviolet, infrared, visible

3.   Heat is energy in the process of being transferred from:

    a.   low pressure to high pressure
    b.   cold objects to hot objects
    c.   high pressure to low pressure
    d.   hot objects to cold objects
    e.   regions of low density toward regions of high density

4.   The rate at which radiant energy is emitted by a body:

    a.   increases with decreasing temperature
    b.   increases with increasing temperature
    c.   does not depend on the body's temperature

5.   If the earth had no atmospheric greenhouse effect, the average surface temperature would be:

    a.   lower than at present
    b.   higher than at present
    c.   the same as it is now

6.   Which below is *not* a primary reason for the seasons in the middle latitudes of the Northern Hemisphere?

    a.   the closeness of the earth to the sun
    b.   the angle at which sunlight reaches the earth
    c.   the length of daylight hours

7.   The moon's surface can only cool by (hint: the moon has no atmosphere):

    a.   convection
    b.   conduction
    c.   radiation

8.   The earth's atmospheric greenhouse effect is produced mainly by water vapor and carbon dioxide absorbing and re-emitting:

    a.   visible radiation
    b.   infrared radiation
    c.   ultraviolet radiation

9. Annually, polar regions lose more heat energy than they receive, yet they are prevented from becoming progressively colder each year mainly by the:

    a. absorption of heat by snow and ice surfaces
    b. conduction of heat through the interior of the earth
    c. storage of heat in the soil beneath the snow cover
    d. circulation of heat by the atmosphere and oceans
    e. release of sensible heat to the atmosphere when the polar ice melts

10. The most important reason why summers in the Southern Hemisphere are not warmer than summers in the Northern Hemisphere is that:

    a. a greater percentage of the Southern Hemisphere is covered with water
    b. the earth is farther from the sun during the Southern Hemisphere summer
    c. the Southern Hemisphere is cloudier during its summer

True–False

_____ 1. A degree Fahrenheit is larger than a degree Celsius.

_____ 2. Sinking air always warms and rising air always cools.

_____ 3. The earth comes closer to the sun in January than it does in July.

_____ 4. Incoming sunlight in middle latitudes is less in winter than in summer partly because the sun's rays slant more and spread their energy over a larger area.

_____ 5. An air temperature of 0°K would be the same as an air temperature of 0°C.

_____ 6. In the middle latitudes of the Northern Hemisphere, between Christmas and New Year's, the length of daylight increases each day.

_____ 7. The sun's radiation is also referred to as shortwave radiation.

_____ 8. Clouds are poor absorbers and emitters of infrared radiation.

_____ 9. Only selective absorbers in the atmosphere emit radiation.

_____ 10. The term "latent" means hidden.

_____ 11. On the average, about 50 percent of the solar radiation that strikes the outer atmosphere eventually reaches the earth's surface.

_____ 12. An ultraviolet photon carries more energy than an infrared photon.

_____ 13. Air is a poor conductor of heat.

_____ 14. The earth's atmosphere behaves as a blackbody.

_____ 15. On the average, each year the earth-atmosphere system sends off into space just as much energy as it receives.

_____ 16. The earth's radiative equilibrium temperature is lower than the earth's observed average surface temperature.

_____   17. An object with a high albedo appears brighter than an object with a low albedo.

_____   18. The process of condensation releases sensible heat into the environment.

Additional Questions

1.   In the spaces provided in Figure 1, convert the temperature from degrees Fahrenheit into degrees Celsius. (Hint: conversion factors are found in Appendix A of your textbook.)

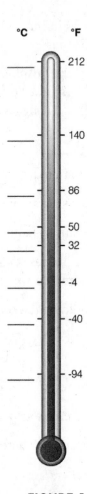

°C   **°F**

_____   — 212

_____   — 140

_____   — 86

_____   — 50
_____   — 32

_____   — -4

_____   — -40

_____   — -94

FIGURE 1

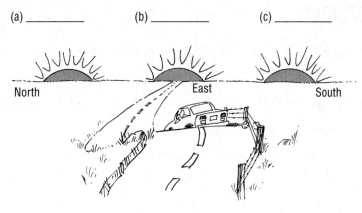

(a) _____   (b) _____   (c) _____

North   East   South

FIGURE 2

2. The diagram above (Fig. 2) represents sunrise on three different days. One day is the autumnal equinox (Sept. 22), one is the Summer solstice (June 21), and the third is the winter solstice (Dec. 21).

    a. In the space above each sun, place the proper date.

    b. Look back through old newspapers (and court records if available) to see if there is a propensity for accidents (both auto and pedestrian) near sunrise around the autumnal equinox in your area. If there is, explain why with the aid of Figure 2.

Additional Readings

"All That's Best of Dark and Bright" by Craig F. Bohren, *Weatherwise*, Vol. 43, No. 3 (June 1990), p. 160.

"The Greenhouse Effect Revisited" by Craig F. Bohren, *Weatherwise*, Vol. 42, No. 1 (February 1989), p. 50.

"Trace Gases, CO_2, Climate and the Greenhouse Effect" by Gordon J. Aubrecht, II, *Physics Teacher*, Vol. 26, No. 3 (March 1988), p. 145.

"The Sun Also Surprises" by Doug Addison, *Weatherwise*, Vol. 46, No. 6 (December 1993), p. 32.

"The Mystery of Disappearing Heat" by Richard Williams, *Weatherwise*, Vol. 49, No. 4 (August/September 1996), p. 28.

"When the Heavens Dance" by Elinor De Wire, *Weatherwise*, Vol. 48, No. 6 (December 1995/January 1996), p. 18.

"The Awesome Aurora" by Gregory J. Byrne and Susan K. Runco, *Weatherwise*, Vol. 52, No. 3 (May/June 1999), p. 38.

"Seeing the Light" by Karen Wright, *Discover*, Vol. 21, No. 7 (July 2000), p. 50.

"Auroras: Earth's Grand Show of Lights" by Kenny Taylor, *National Geographic*, Vol. 200, No. 5 (November 2001), p. 48.

"A Midwinter Night's Dream," Michael Klensch, *Weatherwise*, Vol. 56, No. 1 (January/February 2003), p. 32.

"The Origin of the Solar Wind," by Richard Woo and Shadia Rifai Habbal, *Scientific American*, Vol. 90, No. 6 (November/December 2002), p. 532.

"A Stormy Star," by Curt Suplee, *National Geographic*, Vol. 206, No. 1 (July 2004), p. 2.

"Here Comes the Sun," by Dana Mackenzie, *Discover*, Vol. 25, No. 5 (May 2004), p. 62.

"The Global Light Show," by Jim Reed, *Weatherwise*, Vol. 57, No. 4 (July/August 2004), p. 32.

Answers

Matching

1. g
2. d
3. l
4. b
5. p
6. i
7. n
8. e
9. j
10. m
11. a
12. k
13. q
14. c
15. h
16. o
17. f

Additional Matching

1. c
2. a
3. f
4. e
5. b
6. d

Fill in the Blank

1. kinetic energy
2. reflected
3. blackbody
4. none
5. albedo
6. water vapor and carbon dioxide
7. infrared radiation
8. 30
9. summer
10. visible, infrared
11. water vapor
12. two
13. autumnal equinox
14. window
15. compression

Multiple Choice

1. b
2. b
3. d
4. b
5. a
6. a
7. c
8. b
9. d
10. a

True–False

1. F
2. T
3. T
4. T
5. F
6. T
7. T
8. F
9. F
10. T
11. T
12. T
13. T
14. F
15. T
16. T
17. T
18. T

Additional Questions

1. $-94°F = -70°C$
   $-40°F = -40°C$
   $-4°F = -20°C$
   $32°F = 0°C$

   $50°F = 10°C$
   $86°F = 30°C$
   $140°F = 60°C$
   $212°F = 100°C$

2. a. (a) summer solstice
        (b) autumnal equinox
        (c) winter solstice

    b. On the autumnal equinox, looking east at sunrise would find the sun directly in the eyes of a driver.

Air Temperature

3

The first part of Chapter Three describes how the air temperature varies near the earth's surface during the course of a day. Here we learn that the daily variation in temperature is controlled mainly by the input of solar energy and the output of energy from the surface. When energy input exceeds output, the air temperature rises; when output exceeds input, it falls. At night, the ground cools more quickly than the air above and a radiation temperature inversion often forms. The chapter then looks at the different methods used to protect sensitive crops from the cold surface air.

After considering how and why temperatures vary on a global scale, the chapter examines the significance of temperature variations in terms of practical applications to everyday living. Here we see that temperature information can influence our lives in many ways, from determining what clothes we take on a trip to providing us with critical information for energy-use predictions and agricultural planning. The chapter concludes by examining the wind-chill index and the different instruments that measure air temperature.

Some important concepts and facts of this chapter:

1. During the day, the surface of the earth and the air above it will continue to warm as long as incoming energy (mainly sunlight) exceeds outgoing heat energy from the surface.

2. During the night, the earth's surface cools by giving up more infrared radiation than it receives—a process called radiational cooling.

3. The lowest temperature during the night and early morning hours are usually observed at the earth's surface.

4. When the coldest air is at the earth's surface, the air above is warmer and a radiation inversion exists.

5. The coldest nights occur (typically in winter) when the air is clear, calm, and dry.

6. Most of the methods used to protect sensitive crops from the cold surface air either heat the surface air or mix it with the warmer air above.

7. Large bodies of water warm more slowly than adjacent land areas, and cool more slowly as well.

8.   During the summer, humid climates tend to have lower daytime maximum temperatures and higher nighttime minimum temperatures than do drier climates.

9.   Even though two cities may have similar average annual temperatures, the range and extreme of their temperatures can vary greatly.

10. Wind-chill is a measure of the cooling effect of the wind on the human body. It *does not* relate to inanimate objects such as water pipes, etc.

11. All thermometers must, in some way, be shielded from the sun when measuring air temperature, otherwise their readings will be inaccurate.

Self Tests

....

Match the Following

_____ 1. Warmer hillsides that are less likely to experience freezing conditions

_____ 2. A recording thermometer

_____ 3. The main factors that cause variations in temperature from one place to another

_____ 4. Used as an index for fuel consumption

_____ 5. Thermometers with a small constriction just above the bulb

_____ 6. The rapid lowering of human body temperature can produce this

_____ 7. Obtains air temperature by measuring emitted infrared energy

_____ 8. Thermometer most likely to contain alcohol

_____ 9. These mix the air next to the ground by setting up convection currents

_____ 10. Used as a guide to planting and for determining the approximate date when crops will be ready for harvesting

_____ 11. Protects instruments from the weather elements

a. heating-degree day

b. radiometer

c. thermograph

d. instrument shelter

e. orchard heaters

f. thermal belts

g. maximum thermometer

h. growing-degree day

i. controls of temperature

j. minimum thermometer

k. hypothermia

Fill in the Blank

1. The difference between the highest and lowest temperature for any given day is called the daily _____ of temperature.

2. The amount of heat needed to raise the temperature of one gram of a substance by one degree Celsius is the _____ heat.

3. On a clear, calm night, the ground and air above cool mainly by this process: _____ _____.

4. The difference between the average temperature of the warmest and coldest months is called the annual _____ of temperature.

5. Lines of equal temperature are called _____.

6. A measured increase in air temperature just above the ground is known as a _____ _____.

7. List the four necessary conditions for the development of a strong radiation inversion: _____, _____, _____, _____.

Multiple Choice

1. In clear weather the air next to the ground is usually _____ than the air above during the night, and _____ than the air above during the day.

    a. colder, warmer
    b. colder, colder
    c. warmer, colder
    d. warmer, warmer

2. Near the earth's surface, when outgoing infrared energy *exceeds* incoming solar energy, the air temperature:

    a. increases
    b. decreases
    c. does not change

3. The wind-chill index:

    a. determines how low the air temperature will be on any given day
    b. tells farmers when to protect crops from a freeze
    c. takes into account humidity and air temperature in expressing the current air temperature
    d. indicates the temperature at which water freezes on exposed skin
    e. relates body heat loss with wind to an equivalent temperature with no wind

4.   The primary cause of a radiation inversion is:

    a.   infrared radiation absorbed by the earth's surface
    b.   infrared radiation absorbed by the atmosphere and clouds
    c.   infrared radiation emitted by the earth's surface
    d.   solar radiation absorbed by the earth's surface
    e.   solar radiation reflected by the earth's surface

5.   The largest annual range of temperatures is found:

    a.   in polar latitudes over land
    b.   in polar latitudes over water
    c.   at the equator
    d.   in middle latitudes near large bodies of water
    e.   in the northern Central Plains of the United States

6.   Which below is *not* a liquid-in-glass thermometer?

    a.   maximum thermometer
    b.   minimum thermometer
    c.   bimetallic thermometer

7.   If tonight's air temperature is going to drop into the middle 20s (°F) and a fairly stiff wind is predicted, probably the best way to protect an orchard against a hard freeze is to: (cost is not a factor)

    a.   use wind machines
    b.   use helicopters
    c.   put orchard heaters to work
    d.   sprinkle the trees with water
    e.   pray for clouds

8.   Suppose yesterday morning you noticed ice crystals (frost) on the ground, yet the minimum temperature reported in the newspaper was only 35°F. The *most* likely reason for this apparent discrepancy is that:

    a.   the temperature reading was taken in an instrument shelter more than 5 feet above the ground
    b.   the thermometer was in error
    c.   the newspaper reported the wrong temperature
    d.   the thermometer was read before the minimum temperature was reached for the day
    e.   the thermometer was read incorrectly

9.   An important reason for the large daily temperature range over deserts:

    a.   the light colored sand radiates heat very rapidly at night
    b.   dry air is a very poor heat conductor
    c.   there is little water vapor in the air to absorb and radiate infrared radiation
    d.   the ozone content of desert air is very low
    e.   free convection cells are unable to form over the hot desert ground

True–False

_____   1.   Hypothermia is most common in cold, wet weather.

_____   2.   The greatest variation in daily temperature normally occurs at the ground.

_____   3.   One reason why water warms and cools more slowly than land is because water has a lower specific heat.

_____   4.   An ordinary liquid-in-glass thermometer held in direct sunlight will always indicate a temperature higher than the true air temperature.

_____   5.   In most areas, the warmest time of the day about 5 feet above the ground occurs around noon.

_____   6.   During the summer, humid regions typically have lower daily temperature ranges and lower maximum temperatures than do drier regions.

_____   7.   During the afternoon, the greatest temperature difference between the surface air and the air several feet above occurs on clear, calm afternoons.

_____   8.   If two cities have the same mean annual temperature, then their temperatures throughout the year must be quite similar.

_____   9.   When the air temperature is 35°F and the wind is blowing at 30 miles per hour, the wind-chill index will be below freezing and exposed water pipes will likely freeze.

_____   10.   Radiation temperature inversions are best developed in the early morning just before sunrise.

_____   11.   If you travel from Dallas, Texas to St. Paul, Minnesota during July, you are more likely to experience greater temperature variations than if you made the same trip in January.

_____   12.   In a hilly region, the best place to plant crops that are sensitive to low temperatures is on the valley floor.

_____   13.   Clear, calm nights are usually cooler than cloudy, calm nights.

Additional Questions

1.   The adjacent diagram (Fig. 1) is a section of a minimum thermometer.

   a.   the current air temperature is _____°F

   b.   the minimum temperature is _____°F

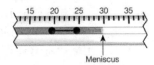

Meniscus

FIGURE 1

2.   How many heating degree-days would there be in a city when the maximum temperature is 40°F and the minimum temperature is 20°F? (Assume a base temperature of 65°F.)

3.   How many cooling degree-days would there be in a city when the maximum temperature is 95°F and the minimum temperature is 75°F? (Assume a base temperature of 65°F.)

4.   Suppose corn is planted in Maryland on March 1. If this variety of corn needs 2400 growing degree-days before it can be picked, and if the mean daily temperature for March through August is 65°F, in about how many days would the corn be ready to pick? On about what date would this be? (Assume a base temperature of 50°F.)

Pick corn in about _____ days.

The date of picking will be about _____.

5.   In the wind-chill tables of your text (p. 71), determine the approximate wind chill under the following conditions:

   a.   air temperature −10°F, wind speed 20 mi/hr   Answer _____

   b.   air temperature −20°C, wind speed 30 km/hr   Answer _____

Additional Readings

"Weather Records: The Case of the Bennett, Colorado, Maximum Temperature" by Thomas W. Bettge, *Weatherwise*, Vol. 38, No. 2 (April 1985), p. 95.

"Windchill: The 'Burr' Index" by Dennis M. Driscoll, *Weatherwise*, Vol. 40, No. 6 (December 1987), p. 321.

"Degree Days: Heating and Cooling by the Numbers" by J. Murray Mitchell, *Weatherwise*, Vol. 40, No. 6 (December 1987), p. 334.

"Tilting at Wind Chills" by Steve Horstmeyer, *Weatherwise*, Vol. 48, No. 5 (October/November 1995), p. 24.

"Cold Rush" by Robert Hensen, *Weatherwise*, Vol. 55, No. 1 (January/February 2002), p. 14.

"Blueberries on Ice," Paul Lyrene, *Weatherwise*, Vol. 57, No. 3 (May/June 2004), p. 20.

"A Stationary Danger," by Jan Null, *Weatherwise*, Vol. 58, No. 4 (July/August 2005), p. 52.

Answers

Matching

1. f	4. a	7. b	10. h
2. c	5. g	8. j	11. d
3. i	6. k	9. e	

Fill in the Blank

1. range
2. specific
3. radiational cooling
4. range
5. isotherms

6. radiation inversion
7. a. clear
   b. calm
   c. dry (little water vapor)
   d. long winter night

Multiple Choice

1. a	4. c	7. d
2. b	5. a	8. a
3. e	6. c	9. c

True–False

1. T	5. F	9. F	13. T
2. T	6. T	10. T	
3. F	7. T	11. F	
4. T	8. F	12. F	

Additional Questions

1. a. about 30°
   b. about 25°
2. 35 heating degree-days
3. 20 cooling degree-days

4. about 160 days; the date for picking would be around August 7
5. a. −35°F
   b. −32.6°C

Humidity, Condensation, and Clouds

Chapter Four begins by examining the concepts of evaporation, condensation, and saturation. It then looks at the many ways of describing humidity. Here we learn that although relative humidity is the most common way to describe atmospheric moisture, it is also the most misunderstood. After a discussion on relative humidity and human discomfort, we learn that a good indicator of the air's actual water vapor content is the dew-point temperature. The chapter continues with a discussion on the formation of dew, frost, and fog. The end of the chapter describes how clouds are classified. This section helps with the identification of clouds by providing you with many visual clues.

Some important concepts and facts of this chapter:

1. Saturation exists when the number of water molecules evaporating from a liquid equals the number condensing.

2. In our atmosphere, condensation occurs primarily when the air is cooled.

3. Condensation nuclei are important in the atmosphere because they serve as surfaces on which water vapor condenses.

4. The air's actual (water) vapor pressure is an indication of the air's water vapor content.

5. Relative humidity, expressed as a percent, does not tell us how much water vapor is actually in the air, rather it tells us how close the air is to being saturated.

6. The dew-point temperature is a good indicator of the air's water vapor content. High dew points indicate high water vapor content and vice versa.

7. When the air temperature and dew point are close together, the relative humidity is high; when they are far apart, the relative humidity is low.

8. High relative humidities in hot weather can make us feel it is hotter than it actually is by retarding the evaporation of perspiration.

9. Dew, frost, and frozen dew form when objects on the surface cool below the air's dew-point temperature.

10. Fog can form as the air cools, or as water evaporates and mixes with drier air.

11. Clouds are usually divided into four main groups: high, middle, low, and clouds with vertical development.

12. A removable color cloud chart is included at the back of your textbook. Take it out and use it while observing the sky.

Self Tests

. . . .

Match the Following

_____ 1. The maximum pressure that water vapor would exert if the air were saturated

_____ 2. These particles serve as surfaces on which water vapor may condense

_____ 3. When fog "burns off" it does this

_____ 4. Combines air temperature with relative humidity to determine an apparent temperature

_____ 5. Beads of water that have condensed onto objects near the surface

_____ 6. Uses wet-bulb and dry-bulb temperatures to obtain relative humidity

_____ 7. Fog that most commonly forms on clear nights, with light or calm winds

_____ 8. Measures relative humidity with human hairs

_____ 9. A tiny liquid drop of dew that freezes when the air temperature drops below freezing

_____ 10. Fog that forms when the wind blows relatively warm air over a colder surface

_____ 11. These clouds appear as baglike sacks hanging beneath a cloud

_____ 12. Fog that forms as moist air flows upward along an elevated surface

_____ 13. The lowest temperature that can be obtained by evaporating water into the air

_____ 14. Steam fog and arctic sea smoke are a form of this type of fog

a. Heat Index

b. radiation fog

c. saturation vapor pressure

d. evaporation-mixing fog

e. mammatus

f. sling psychrometer

g. evaporates

h. frozen dew

i. wet-bulb temperature

j. condensation nuclei

k. advection fog

l. hair hygrometer

m. upslope fog

n. dew

Additional Matching (Clouds)

(Some answers may be used more than once.)

_____	1.	A "mackerel sky" describes this cloud
_____	2.	A low, lumpy cloud layer that appears in rows, patches, or rounded masses
_____	3.	A towering cloud that has not fully developed into a thunderstorm
_____	4.	Hail is usually associated with this cloud type
_____	5.	The sun or moon are dimly visible or appear watery through this gray, sheetlike cloud
_____	6.	A halo around the sun or moon often identifies the presence of this cloud
_____	7.	Wispy, high clouds
_____	8.	Light or moderate but steady precipitation that covers a broad area is most often associated with this cloud
_____	9.	This cloud's elements (puffs) should be about the size of your thumbnail when your hand is extended to arm's length
_____	10.	Lightning and thunder are associated with this cloud
_____	11.	A cloud of vertical development that resembles a small piece of floating cotton
_____	12.	A middle cloud that occasionally forms in parallel waves or bands
_____	13.	The cloud with the smallest elements or puffs as viewed from the surface
_____	14.	A low, uniform, grayish cloud, whose precipitation is most commonly drizzle
_____	15.	This cloud's elements (puffs) should be about the size of your fist when your hand is extended to arm's length
_____	16.	When fog lifts above the surface, it forms this gray, sheet-like cloud
_____	17.	Cloud with the greatest vertical growth

a. cirrus

b. cirrostratus

c. cirrocumulus

d. altostratus

e. altocumulus

f. nimbostratus

g. stratus

h. stratocumulus

i. cumulus

j. cumulus congestus

k. cumulonimbus

Fill in the Blanks

1. The process of water changing from a liquid to a vapor is called _____.

2. Cirrus clouds are composed primarily of _____ _____.

3. _____ _____ can be described as the percent of water vapor in the air compared to that required for saturation.

4. The cooling of the ground to produce dew and frost is mainly the result of _____ cooling.

5. On most days the relative humidity reaches its highest value when the air temperature reaches its _____ value.

6. Clouds with a lens shape that often form over and downwind of mountains are called _____ clouds.

7. The temperature to which air must be cooled for saturation to occur is called the _____ _____.

8. _____ forms when water vapor changes directly into ice without becoming a liquid first.

9. A cloudlike stream seen forming behind a jet aircraft is called a(n) _____.

10. The circulation of water within the atmosphere is called the _____ cycle.

11. Instruments that measure humidity are called _____.

Multiple Choice

1. Which is the *best* indicator of the actual amount of water vapor in the air?

    a. air temperature
    b. dew-point temperature
    c. relative humidity
    d. wet-bulb temperature
    e. saturation vapor pressure

2. If the air temperature remains constant, evaporating water into the air will _____ the dew-point temperature and _____ the relative humidity.

    a. decrease, decrease
    b. decrease, increase
    c. increase, decrease
    d. increase, increase

3. If you are standing outside and notice that the sky is covered with a high, white layered cloud, and you look at the ground and observe your shadow, you may conclude that the cloud overhead is:

    a. altostratus
    b. cirrostratus
    c. stratus
    d. nimbostratus
    e. stratocumulus

4. When the air temperature increases, the saturation vapor pressure:

    a. increases
    b. decreases
    c. does not change

5. If you are interested in the air's water vapor density, then you would be interested in the:

    a. specific humidity
    b. mixing ratio
    c. absolute humidity
    d. relative humidity

6. Of the different types of fog listed below, which one does *not* necessarily form in air that is cooling?

    a. advection fog
    b. radiation fog
    c. upslope fog
    d. evaporation-mixing fog

7. The highest clouds in our atmosphere are called:

    a. cirrus
    b. altocumulus
    c. noctilucent
    d. cumulonimbus
    e. cumulus congestus

8. Polar air is considered "dry" because the dew-point temperatures are often quite low. However, the relative humidity of this cold, polar air is usually high because:

    a. low dew points indicate that the relative humidity must be high
    b. low air temperatures indicate that the relative humidity must be high
    c. the air temperature and the dew point are fairly close together

9. A cloud that sometimes resembles a silken scarf capping the top of a developing cumulus cloud is the:

    a. nacreous cloud
    b. noctilucent cloud
    c. pileus cloud
    d. mammatus cloud
    e. scud

10. The most common type of fog that forms over the Pacific coastal waters of North America is:

    a. advection fog
    b. radiation fog
    c. evaporation-mixing fog
    d. upslope fog

11. Which of the clouds listed below is *least* likely to produce precipitation that reaches the ground?

    a. nimbostratus
    b. cumulus congestus
    c. cumulonimbus
    d. stratus
    e. cirrocumulus

12. As the air temperature *decreases*, the likelihood of condensation occurring:

    a. increases
    b. decreases
    c. does not increase or decrease

13. These clouds form in the stratosphere and are also called mother-of-pearl clouds:

    a. cumulonimbus clouds
    b. noctilucent clouds
    c. nacreous clouds
    d. nimbostratus clouds
    e. pileus clouds

14. Which condition below would *best* describe supersaturated air:

    a. relative humidity is zero (0) percent.
    b. relative humidity is 50 percent.
    c. relative humidity is 100 percent.
    d. relative humidity is 110 percent.

True–False

_____   1.   The relative humidity is a measure of the air's actual water vapor content.

_____   2.   All other factors being equal, increasing wind speed enhances evaporation.

_____   3.   On a hot, humid day, a good measure of how cool the human skin can become is the wet-bulb temperature.

_____   4.   Fog can be composed of ice crystals.

_____   5.   Dew is more likely to form on clear, windy nights.

_____   6.   A cloud that forms in descending air is the mammatus.

_____   7.   Relative humidity is always given as a percent.

_____   8.   The process by which water changes from a vapor to a solid is called deposition.

_____   9.   In middle latitudes, high clouds are typically observed below an altitude of 20,000 ft or 6000 m.

_____ 10.   Valleys are more susceptible to radiation fog than are hilltops.

_____ 11.   Advection fog often forms as warm rain falls into a cold layer of surface air.

_____ 12.   If you turn on your oven and open its door, the increase in air temperature would raise the relative humidity inside your home.

_____ 13.   Near the earth's surface at the same temperature and level in the atmosphere, warm humid air is less dense than warm dry air.

_____ 14.   When the air is saturated, an increase in air temperature will cause condensation to occur.

_____ 15.   Another name for "luminous night cloud" is noctilucent cloud.

_____ 16.   Suppose the air temperature inside your home is 78°F and you lower it to 68°F. As long as the moisture content of the air inside does not change, the relative humidity should increase.

Additional Questions

1.   List the two most important factors that cause the relative humidity of air to change.

   a.

   b.

2. a. Using Fig. 4.5, p. 83, in your textbook, determine the saturation vapor pressure for the following air temperatures:

Air Temperature	Saturation Vapor Pressure (millibars)
0°C	_____
10°C	_____
20°C	_____
30°C	_____

b. Suppose the outside air temperature is 30°C (86°F). You know from (a) above that the saturation vapor pressure is 42 millibars. If the actual vapor pressure of the air is 21 millibars, use the following relative humidity formula to determine the relative humidity of the air:

$$RH = \frac{\text{actual vapor pressure}}{\text{saturation vapor pressure}} \times 100 \text{ percent}$$

Relative humidity = _____%

3.   In the space provided, write in the name of the cloud that is depicted in the illustration. (Additional photographs of clouds are found in the color cloud chart bound at the back of your textbook.)

FIGURE A

a.   _____

(Hint: Light continuous snow is falling from this cloud.)

FIGURE B

b.   _____

FIGURE C

c.   _____

FIGURE D

(Hint: This is classified as a low cloud.)

d. _____

FIGURE E

(Hint: This is a high cloud.)

e. _____

FIGURE F

f. _____

FIGURE G

g. _____

(Hint: Sun dimly visible.)

FIGURE H

h. _____

(Hint: This is a high cloud.)

FIGURE I

i. _____

FIGURE J

j. _____

FIGURE K

k. _____

4. a. In Fig. 4.10, p. 89, in your textbook (the Heat Index chart), determine the apparent temperature under the following conditions:

Air Temperature	Relative Humidity	Apparent Temperature (°F)
80°F	90%	_____
90°F	70%	_____
100°F	50%	_____
110°F	30%	_____

Additional Readings

"How Humid is Humid" by David E. Siskind, *Weatherwise*, Vol. 44, No. 3 (June 1991), p. 24.

"Water Vapor Mysticism" by Craig Bohren, *Weatherwise*, Vol. 43, No. 2 (April 1990), p. 97.

"Sugar and Spice: The Dirty Wet-Bulb Temperature" by Craig F. Bohren, *Weatherwise*, Vol. 39, No. 1 (February 1986), p. 46.

"Building a Dew Point Hygrometer" by Hampton W. Shirer, *Weatherwise*, Vol. 39, No. 2 (June 1986), p. 160.

"Fog on Trial" by Stanley David Gedzelman, *Weatherwise*, Vol. 44, No. 2 (April 1991), p. 14.

"Five Faces of Freezing" by Craig F. Bohren, *Weatherwise*, Vol. 24, No. 6 (December 1989), p. 315.

"What Your Windshield Shows about the Clouds" by Richard Williams, *Weatherwise*, Vol. 40, No. 5 (October 1987), p. 251.

"An Essay on Dew" by Craig Bohren, *Weatherwise*, Vol. 41, No. 4 (August 1988), p. 226.

A Field Guide to the Atmosphere by Vincent J. Schaefer and John A. Day; Houghton Mifflin, Boston, 1981.

"Fog on the U.S. West Coast: A Review" by D. F. Leipper, *Bulletin of the American Meteorological Society*, Vol. 75, No. 2 (February 1994), p. 229.

"All That Glistens Isn't Dew" by Craig F. Bohren, *Weatherwise*, Vol. 43, No. 5 (October 1990), p. 284.

"Cloud Classification Before Luke Howard" by Stanley David Gedzelman, *Bulletin of the American Meteorological Society*, Vol. 70, No. 4 (April 1989), p. 381.

Peterson's First Guide to Clouds and Weather by John A. Day and Vincent J. Schaefer; Houghton Mifflin, Boston, 1991.

"An Intimate Look at Clouds" by Brooks Martner, *Weatherwise*, Vol. 49, No. 3 (June/July 1996), p. 20.

"Noctilucent Clouds" by Gary Thomas and Jay Brausch, *Weatherwise*, Vol. 50, No. 3 (June/July 1997), p. 32.

"A Focus on Frost" by Mark Schneider, *Weatherwise*, Vol. 52, No. 6 (November/December 1999), p. 36.

"Cumulus Patterns" by Gregory J. Byrne, *Weatherwise*, Vol. 53, No. 3 (May/June 2000), p. 30.

"Angel Hair Cirrus" by Stanley David Gedzelman, *Weatherwise*, Vol. 53, No. 4 (July/August 2000), p. 20.

"Sentinels in the Sky" by Jeff Rosenfeld, *Weatherwise*, Vol. 53, No. 1 (January/February 2000), p. 24.

"A Cloud by any other name ...," Stanley David Gedzelman, *Weatherwise*, Vol. 56, No. 6 (November/December 2003), p. 24.

"Living with Heat," Mace Bentley and Steve Horstmeyer, *Weatherwise*, Vol. 57, No. 1 (January/February 2004), p. 16.

"Foggy Landings," Douglas Morris, *Weatherwise*, Vol. 57, No. 1 (January/February 2004), p. 38.

"In Search of the Morning Glory," by Gavin Pretor-Pinney, *Weatherwise*, Vol. 58, No. 1 (January/February 2006), p. 20.

"A Humid Relationship," by Stephen Smulski, *Weatherwise*, Vol. 59, No. 3 (May/June 2006), p. 38.

"A Stationary Danger," by Jan Null, *Weatherwise*, Vol. 58, No. 4 (July/August 2005), p. 52.

"Silent Sizzle: Tracking Deadly Heat," by Jan Null, *Weatherwise*, Vol. 59, No. 5, (September/October 2006), p. 36.

Answers

Matching

1.	c	5.	n	9.	h	13.	i
2.	j	6.	f	10.	k	14.	d
3.	g	7.	b	11.	e		
4.	a	8.	l	12.	m		

Additional Matching (Clouds)

1.	c	6.	b	11.	i	16.	g
2.	h	7.	a	12.	e	17.	k
3.	j	8.	f	13.	c		
4.	k	9.	e	14.	g		
5.	d	10.	k	15.	h		

Fill in the Blank

1. evaporation
2. ice crystals
3. relative humidity
4. radiational
5. lowest or minimum
6. lenticular
7. dew-point temperature or dew point
8. frost
9. contrail or condensation trail
10. hydrologic
11. hygrometers

Multiple Choice

1.	b	5.	c	9.	c	13.	c
2.	d	6.	d	10.	a	14.	d
3.	b	7.	c	11.	e		
4.	a	8.	c	12.	a		

True–False

1.	F	5.	F	9.	F	13.	T
2.	T	6.	T	10.	T	14.	F
3.	T	7.	T	11.	F	15.	T
4.	T	8.	T	12.	F	16.	T

Additional Questions

1. a. change in air temperature
   b. change in water vapor content

2. a.

Air Temperature	Saturation Vapor Pressure
0°C	about 6 millibars
10°C	about 12 millibars
20°C	about 23 millibars
30°C	about 42 millibars

   b. Relative humidity $= \dfrac{\text{actual vapor pressure}}{\text{saturation vapor pressure}} \times 100\%$

$$RH = \frac{21}{42} \times 100\% = 50\%$$

3. a. cumulonimbus
   b. nimbostratus
   c. lenticular
   d. stratocumulus
   e. cirrostratus
   f. cumulus
   g. cirrus
   h. altostratus
   i. cirrocumulus
   j. altocumulus
   k. mammatus

4. a.

Air Temperature	Relative Humidity	Apparent Temperature
80°F	90%	88°F
90°F	70%	106°F
100°F	50%	120°F
110°F	30%	123°F

Cloud Development and Precipitation

5

Chapter Five ties together the concepts of stability, cloud formation, and precipitation. The first part of the chapter looks at stable and unstable air. Here we learn that stable air tends to resist vertical motions, while unstable air tends to favor vertical air motions. After describing the causes of stable and unstable atmospheres, the chapter discusses the different ways clouds form. At this point we see that instability generated by surface heating can produce cumulus clouds, and that topographic barriers can greatly influence cloud development. The chapter then examines precipitation processes. Here we see how cloud particles are able to grow large enough to fall as rain or snow and how falling raindrops can be changed into other forms of precipitation, such as sleet and freezing rain. The last section describes how precipitation is measured.

Some important concepts and facts of this chapter:

1. Stable air tends to resist upward vertical motions. Consequently clouds that form in a stable atmosphere tend to spread horizontally and have a stratified appearance, such as nimbostratus and altostratus.

2. The atmosphere is absolutely *stable* when the air at the surface is either cooler than the air aloft (an inversion), or the temperature difference between the warmer surface air and the cooler air aloft is not very great.

3. The atmosphere can be made more stable by cooling the surface air, warming the air aloft, or by causing air to sink (subside) over a vast area.

4. Unstable air tends to enhance vertical air motion. Consequently, clouds that form in an unstable atmosphere tend to develop vertically and sometimes build to great heights.

5. The atmosphere is absolutely *unstable* when the surface air is much warmer than the air aloft.

6. The atmosphere can be made more unstable by warming the surface air, by cooling the air aloft, or by lifting a layer of air.

7. The majority of clouds form due to surface heating, the convergence (flowing together) of surface air, and forced uplift along topographic barriers and weather fronts.

8. The combining (merging) of tiny cloud droplets into a large droplet is called coalescence.

9. During the ice crystal (Bergeron) process, ice crystals, surrounded by water droplets, grow larger at the expense of the droplets; consequently as ice crystals grow larger, water droplets become smaller.

10. Accretion is the process by which ice crystals or snowflakes grow larger by colliding with supercooled liquid droplets that freeze on contact.

11. It is never too cold to snow.

12. Freezing rain forms when cold raindrops fall through a shallow, subfreezing layer and freeze upon striking the ground or the surface of a cold object.

13. Sleet (ice pellets) forms when cold raindrops, or partially melted snowflakes, fall through a relatively deep subfreezing layer and turn back into solid ice before reaching the surface.

14. Hailstones form only in thunderstorms that have updrafts capable of keeping ice particles suspended within the cloud long enough to acquire further coatings of ice.

Self Tests

Match the Following:

_____	1.	Type of uplift that occurs as air is lifted over a mountain barrier
_____	2.	Drier region observed on the downwind (leeward) side of a mountain range
_____	3.	An aggregate of ice crystals
_____	4.	Injecting a cloud with small particles with the intent to enhance precipitation
_____	5.	The smallest raindrops
_____	6.	The process by which an air parcel expands and cools or compresses and warms with no interchange of heat with its surroundings
_____	7.	The largest form of precipitation
_____	8.	Intermittent, heavy downpour of either rain or snow
_____	9.	Particles in the atmosphere on which ice crystals may grow
_____	10.	These form when snowflakes or ice crystals fall from high cirriform clouds
_____	11.	Brittle, crunchy pieces of snowlike ice that usually fall as a shower from a cumuliform cloud
_____	12.	Rain that falls from a cloud but evaporates before reaching the surface
_____	13.	An intense snow shower
_____	14.	High winds, low temperatures and blowing or falling snow
_____	15.	Measures rainfall intensity and amount

a. seeding

b. virga

c. orographic

d. hail

e. shower

f. rain shadow

g. Doppler radar

h. fallstreaks

i. snow squall

j. drizzle

k. blizzard

l. ice nuclei

m. snow pellets

n. adiabatic

o. snowflake

Fill in the Blank

1.   The merging of cloud droplets by collision is called _____.

2.   An amount of precipitation measured to be less than one hundredth of an inch is called a _____.

3.   If an air parcel is given a small push upward and it continues to move upward on its own accord, the atmosphere is said to be _____.

4.   The rate at which the air temperature changes inside a rising or descending parcel of *unsaturated* air is called the _____ _____ _____.

5.   Cold rain that falls through a rather shallow layer of subfreezing air and freezes upon striking a surface is called _____ _____.

6.   If an air parcel is given a small push upward and it falls back to its original position, the atmosphere is said to be _____.

7.   The growth of a precipitation particle by the collision of an ice crystal or snowflake with a liquid droplet at temperatures below freezing is called _____.

8.   The rate at which the temperature changes inside a rising or descending parcel of *saturated* air is called the _____ _____ _____.

9.   If unsaturated stable air is lifted to a level where it becomes saturated and unstable, this type of instability is called _____ _____.

10.   Light showers of snow that fall intermittently from cumuliform clouds for a short duration are called _____ _____.

11.   A cold raindrop (or partially melted snowflake) that freezes into a pellet of ice in a deep, subfreezing layer of surface air is called _____.

12.   Water droplets that exist at temperatures below freezing are said to be _____.

Multiple Choice

1. Rising saturated air cools at a lesser rate than rising unsaturated air primarily because:

    a. rising saturated air is heavier
    b. rising saturated air is lighter
    c. unsaturated air expands more rapidly
    d. saturated air does not expand
    e. latent heat of condensation is released in a rising parcel of saturated air

2. For least polluted surface air, the best time of the day for a farmer to burn agricultural debris would be:

    a. around sunrise
    b. around sunset
    c. around the time of maximum surface air temperature
    d. about midnight, so no one will see it

3. Which of the following is *not* a way of producing clouds?

    a. lifting air over a topographic barrier
    b. lifting air along a weather front
    c. warming the surface of the earth
    d. convergence (flowing together) of surface air
    e. subsidence

4. Which of the following is *not* considered an important factor in the production of rain by the collision-coalescence process?

    a. the updrafts in the cloud
    b. the number of ice crystals in the cloud
    c. the cloud thickness
    d. the relative size of the droplets
    e. the cloud's liquid water content

5. These two conditions, working together, will make the atmosphere the most *unstable*:

    a. cool the surface and cool the air aloft
    b. cool the surface and warm the air aloft
    c. warm the surface and cool the air aloft
    d. warm the surface and warm the air aloft

6. These two conditions, working together, will make the atmosphere the most *stable*:

    a. cool the surface and cool the air aloft
    b. cool the surface and warm the air aloft
    c. warm the surface and cool the air aloft
    d. warm the surface and warm the air aloft

7. Which of the adjacent illustrations *best* describes the ice-crystal (Bergeron) process of rain formation?

    a.   1
    b.   2
    c.   3
    d.   4

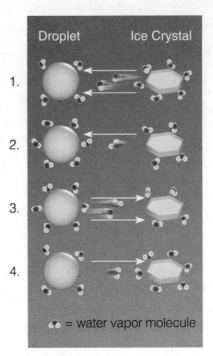

8. The moist adiabatic rate is different from the dry adiabatic rate because:

    a.   saturated air is always unstable
    b.   an unstable air parcel expands more rapidly
    c.   unsaturated air is always stable
    d.   latent heat is released inside a parcel of rising saturated air
    e.   a parcel of saturated air weighs less than a parcel of unsaturated air

9. The two main substances used in cloud seeding are:

    a.   silver iodide and dry ice
    b.   lead iodide and water ice
    c.   cupric sulfide and carbon
    d.   salt particles and dendrite ice crystals

10. Hail forms inside this cloud:

    a.   altocumulus
    b.   cumulonimbus
    c.   stratocumulus
    d.   cumulus humilis
    e.   lenticular

11. Which below best describes the shape of a large falling raindrop about 5 millimeters in diameter?

    a.   spherical
    b.   tear drop
    c.   square
    d.   elongated like that of a cylinder
    e.   slightly elongated and flattened on the bottom

12. If it is raining on one side of the street but not on the other, it is a good bet that the rain is falling from:

    a.   a nimbostratus cloud
    b.   an altocumulus cloud
    c.   a stratus cloud
    d.   an altostratus cloud
    e.   a cumulonimbus cloud

True–False

_____   1.   It is never too cold to snow.

_____   2.   The most common snowflake shape is the dendrite.

_____   3.   Lumpy ice particles, also known as snow pellets, that form in a cloud usually by the process of accretion are called graupel.

_____   4.   Radiational cooling of the surface at night tends to make the lower atmosphere unstable.

_____   5.   Most thunderstorms do not penetrate very far into the stratosphere because the stratosphere is a layer of unstable air.

_____   6.   Air motions tend to be downward around the outside of a cumulus cloud.

_____   7.   The environmental lapse rate represents the rate at which the air temperature is changing in the vertical.

_____   8.   You would use a wooden stick to measure rainfall in the tipping bucket rain gauge.

_____   9.   Much of the rain that falls in middle and high latitudes of North America begins as snow.

_____   10.   During the ice-crystal (Bergeron) process of rain formation, ice crystals grow at the expense of the surrounding water droplets.

_____   11.   Warming the air aloft will tend to make the atmosphere more unstable.

_____   12.   Normally, during the course of a day, the atmosphere near the surface is the most unstable during the afternoon, and the most stable in the early morning.

_____   13.   One would expect a cumulus congestus cloud to be forming in a stable atmosphere.

_____   14.   Freezing rain tends to form a layer of clear ice on the wing of an aircraft.

_____   15.   Radar gathers information about precipitation in clouds by measuring the amount of sunlight reflected off the precipitation.

_____   16.   In a cloud, small droplets tend to fall faster than large droplets.

_____   17.   Cirrus clouds can sometimes naturally seed lower clouds beneath them.

_____   18.   Cloud seeding using silver iodide can work when the air temperature inside the cloud is either above or below freezing.

_____   19.   Large, heavy snowflakes are generally associated with moist air and air temperatures well below freezing.

_____   20.   Doppler radar measures the speed at which falling rain is moving horizontally toward or away from the radar antenna.

Additional Questions

1.   a.   Suppose the environmental lapse rate is 4°F per 1000 feet. If the air temperature at the surface (0 feet) is 70°F, then the air temperature measured by a radiosonde at 4000 feet above the surface would be _____ °F.

     b.   Suppose a parcel of dry air is somehow lifted from the surface (0 feet) to an elevation of 4000 feet. If the temperature inside the air parcel at the surface is 70°F, what would be the temperature of the air inside the parcel at 4000 feet if the rising air parcel remains *unsaturated*? _____ °F (Hint: Use the dry adiabatic rate of 5.5°F per 1000 feet.)

          Look at the temperature of the air at 4000 feet as measured by the radiosonde in (a). Now look at the temperature of the rising air inside the parcel at 4000 feet in (b). Is the air temperature inside the parcel at 4000 feet warmer or colder than the air surrounding it (as measured by radiosonde)? _____ In this situation would the atmosphere be considered stable or unstable? _____ Explain why.

c.   Now, suppose a parcel of *saturated* air is somehow lifted from the surface (0 feet) to an elevation of 4000 feet. If the temperature inside the parcel at the surface is 70°F, what would be the parcel temperature at 4000 feet if the rising air parcel remains *saturated*? _____°F? (Hint: Use a moist adiabatic rate of 3.0°F per 1000 feet.)

What was added to the parcel to make it warmer now at 4000 feet than it was in (b) (when the parcel was lifted and remained unsaturated)?

Look again at the temperature of the air at 4000 feet as measured by a radiosonde in (a). Is the parcel now (after having been lifted at the moist adiabatic rate) colder or warmer than the air surrounding it? _____ Would the atmosphere now be considered stable or unstable? _____ Explain why.

2.   After a large snowstorm the newspaper reports the water equivalent of the precipitation for the following cities to be:

Snowfall

a.   Albany, NY   1.35 inches   _____

b.   Chicago, IL   0.83 inches   _____

c.   Cleveland, OH   0.91 inches   _____

d.   Detroit, MI   0.25 inches   _____

If we assume an average water equivalent ratio of 1:10 for this storm, and that all of the precipitation fell as snow, then how many inches of snow did each city receive?

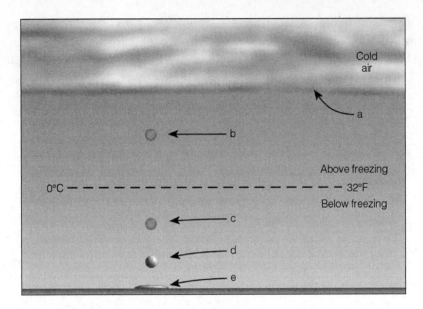

FIGURE 1

3. Answer the following questions that pertain to letters a through e in Figure 1.

a. The thick, relatively low cloud in Figure 1 is stratified and produces light precipitation over a broad area. What is the name of this cloud?

_____

b. At letter b, a cold raindrop is falling toward the surface in above-freezing air. If the raindrop initially formed in the cold part of the cloud as a snowflake, the precipitation probably formed by the _____ process of rain formation.

c. The cold, liquid raindrop at letter c, falling through this region of below-freezing air, is said to be _____.

d. If the cold raindrop freezes at letter d before reaching the ground it will produce precipitation called _____.

e. If the cold raindrop freezes upon striking the surface at letter e, the precipitation is called _____ _____.

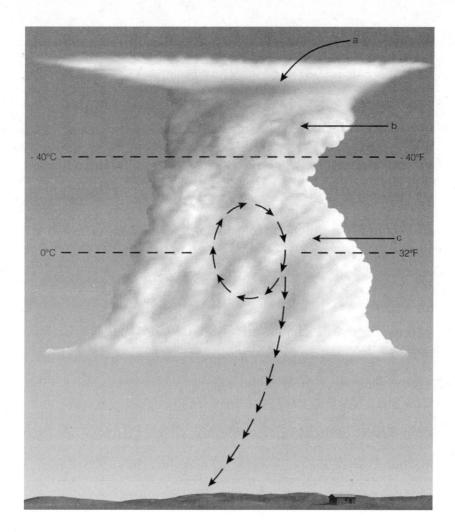

FIGURE 2

4. Answer the following questions that pertain to Figure 2.

    a. The cloud in Figure 2 is what type? _____

    b. What particles would you most likely expect to observe in the region of letter a and letter b in this cloud? _____

    c. What particles would you most likely expect to observe in the region of letter c?

    _____

    d. Suppose a precipitation particle made several up and down journeys, as indicated by the arrows. If the precipitation eventually reaches the surface as ice, it would be called

    _____.

Additional Readings

"How Snow Crystals Grow" by John Hallett, *American Scientist*, Vol. 72, No. 6 (Nov–Dec 1984), p. 582.

"What Becomes of a Winter Snowflake?" by Samuel C. Colbeck, *Weatherwise*, Vol. 38, No. 6 (December 1985), p. 312.

"Of Wet Snow, Slush and Snowballs" by Samuel C. Colbeck, *Weatherwise*, Vol. 39, No. 6 (December 1986), p. 314.

Clouds in a Glass of Beer by Craig F. Bohren; John Wiley and Sons, Inc., New York, 1987. See especially pp. 1–14.

A Field Guide to the Atmosphere by Vincent J. Schaefer and John A. Day; Houghton Mifflin, Boston, 1981. See especially pp. 65–73.

"Hail: The White Plague" by Patrick Hughes and Richard Wood, *Weatherwise*, Vol. 45, No. 2 (April/May 1993), p. 16.

"The Great Sierra Snow Blockade" by Mark McLaughlin, *Weatherwise*, Vol. 45, No. 6 (December 1992/January 1993), p. 20.

"Smells Like Rain" by Robert Henson, *Weatherwise*, Vol. 49, No. 2 (April/May 1996), p. 20.

"Mayday!" by Carolinda Hill, *Weatherwise*, Vol. 49, No. 3 (June/July 1996), p. 25.

"Snow Foolin'" by Nolan J. Doesken and Robert J. Leffler, *Weatherwise*, Vol. 53, No. 1 (January/February 2000), p. 30.

"On Frozen Pond" by Charles Knight, *Weatherwise*, Vol. 51, No. 1 (January/February 1999), p. 35.

"Snow in America" by Bernard Mergen, *Weatherwise*, Vol. 50, No. 6 (December 1997), p. 12.

"Iceberg in the Sky" by Eric Pinder, *Weatherwise*, Vol. 51, No. 6 (November/December 1998), p. 16.

"Seeds of Controversy," Jeff Dick, *Weatherwise*, Vol. 54, No. 6 (November/December 2001), p. 30.

"Balance of Power," Steve Horstmeyer, *Weatherwise*, Vol. 55, No. 1 (January/February 2002), p. 30.

"Designing Snowflakes," Sam Boykin, *Weatherwise*, Vol. 57, No. 1 (January/February 2004), p. 24.

"The Secret Life of Snow," Ivan Amato, *Discover*, Vol. 25, No. 2 (February 2004), p. 56.

"Changing the Weather," by Bridget Coila, *Weatherwise*, Vol. 58, No. 3 (May/June 2005). p. 50.

"Making Rain," by J. Marshall Shephard, *Weatherwise*, Vol. 58, No. 5 (September/October 2005), p. 28.

"Strange Tales of Hail," by Randy Cerveny, et al., *Weatherwise*, Vol. 58, No. 3 (May/June 2005), p. 28.

"Back in Business," by Andrew Freedman, *Weatherwise*, Vol. 59, No. 4 (July/August 2006), p. 24.

"In Search of the Morning Glory," by Gavin Pretor-Inney, *Weatherwise*, Vol. 59, No. 1 (January/February 2006), p. 20.

"The De-Iceman Cometh," by Doug Morris, *Weatherwise*, Vol. 60, No. 1 (January/February 2007), p. 40

Answers

Matching

1.	c	5.	j	9.	l	13.	i
2.	f	6.	n	10.	h	14.	k
3.	o	7.	d	11.	m	15.	g
4.	a	8.	e	12.	b		

Fill in the Blank

1.	coalescence	7.	accretion
2.	trace	8.	moist adiabatic rate
3.	unstable	9.	conditional instability
4.	dry adiabatic rate	10.	snow flurries
5.	freezing rain	11.	sleet
6.	stable	12.	supercooled

Multiple Choice

1.	e	4.	b	7.	c	10.	b
2.	c	5.	c	8.	d	11.	e
3.	e	6.	b	9.	a	12.	e

True–False

1.	T	6.	T	11.	F	16.	F
2.	T	7.	T	12.	T	17.	T
3.	T	8.	F	13.	F	18.	F
4.	F	9.	T	14.	T	19.	F
5.	F	10.	T	15.	F	20.	T

Additional Questions

1. a. 54°F
   b. 48°F
   colder
   Stable, because the rising air is colder than the air surrounding it. The parcel is, therefore, heavier (more dense) than the air around it and would tend to return to its original level.
   c. 58°F
   The air is warmer because as the parcel rose into the atmosphere it was saturated. Cooling of the rising air caused condensation to occur which released vast amounts of latent heat.
   warmer
   Unstable, because the rising air is warmer than the air surrounding it. The parcel is lighter (less dense) than the air around it and would continue to rise on its own accord.

2. a. 13.5 inches
    b. 8.3 inches
    c. 9.1 inches
    d. 2.5 inches
3. a. nimbostratus
    b. ice-crystal or Bergeron
    c. supercooled

    d. sleet
    e. freezing rain
4. a. cumulonimbus
    b. ice crystals or ice particles
    c. mostly liquid cloud droplets, some
        ice
    d. hail

Air Pressure and Winds

6

Chapter Six gives us a broad view of how and why the wind blows. It opens with a section on atmospheric pressure. This is followed by a section that describes surface and upper-air charts. The next section examines the forces that influence atmospheric motions. Here we learn that the wind blows in response to differences in atmospheric pressure and that once air begins to move, the Coriolis force tends to bend it to the right of its intended path in the Northern Hemisphere and to the left in the Southern Hemisphere. The chapter then looks at how the winds blow around pressure systems in the Northern and Southern Hemispheres, both aloft and at the surface. The latter part of the chapter deals with measuring and determining wind direction and wind speed.

Some important concepts and facts of this chapter:

1. Air pressure can be described as a measure of the total mass of air above any level or as the force exerted by air molecules over a given area.

2. Atmospheric pressure decreases most rapidly with elevation in a cold column of air.

3. Cold air aloft is normally associated with low atmospheric pressure, while warm air aloft is associated with high atmospheric pressure.

4. The barometer is the instrument that measures air pressure.

5. The amount of pressure change that occurs over a given horizontal distance is the pressure gradient.

6. Horizontal differences in pressure create a pressure gradient force (PGF). This force causes the wind to blow.

7. On a weather map, closely spaced isobars (or contours) represent a steep pressure gradient, a strong PGF, and high winds, while widely spaced isobars (or contours) represent a gentle pressure gradient, a weak PGF, and light winds.

8. The Coriolis force causes the wind to bend to the right of its path in the Northern Hemisphere and to the left of its path in the Southern Hemisphere.

9. Above middle and high latitudes, the winds on an upper-level chart tend to blow parallel to contour lines (or isobars) in a roughly west-to-east direction in both hemispheres.

10. In the Northern Hemisphere, winds on a surface weather map blow clockwise and outward from the center of a high; counterclockwise and inward toward the center of a low.

11. Sinking air occurs above a surface high-pressure area; rising air above a surface low-pressure area.

12. The prevailing wind is the direction from which the wind blows most frequently during a given time period.

Self Tests

- -

Match the Following

_____	1.	The amount of pressure change that occurs over a given horizontal distance
_____	2.	An apparent force created by the rotation of the earth
_____	3.	Wind aloft that blows in a straight line at a constant speed parallel to the isobars or contours
_____	4.	To correctly monitor horizontal changes in air pressure, this is the most important correction added to the station pressure
_____	5.	Indicates the percent of time the wind blows from different directions
_____	6.	Instrument that employs Doppler radar to obtain a vertical profile of wind speed and wind direction
_____	7.	The force exerted by air molecules over a given area
_____	8.	Lines of equal pressure
_____	9.	The force which balances the vertical pressure gradient force and prevents the atmosphere around the earth from rushing off into space
_____	10.	Instrument that usually consists of three or more cups
_____	11.	An elongated high pressure area
_____	12.	Another name for a large area of low pressure observed in the middle latitudes
_____	13.	The unit of pressure most commonly found on a surface weather map
_____	14.	Surface wind that blows from land to water
_____	15.	A wind-flow pattern with a strong north-south trajectory

a. geostrophic wind

b. mid-latitude cyclone

c. wind profiler

d. anemometer

e. meridional

f. offshore wind

g. pressure gradient

h. gravity

i. wind rose

j. Coriolis force

k. ridge

l. altitude

m. millibar

n. air pressure

o. isobars

Fill in the Blank

1. The force that causes the wind to deflect (bend) to the right of its path in the Northern Hemisphere and to the left in the Southern Hemisphere is the _____ _____.

2. The fundamental laws of motion were formulated by this man: _____ _____.

3. Even though _____ lines are drawn on a 500-millibar chart, they illustrate regions of high-and-low pressure much like isobars do.

4. A recording aneroid barometer can also be called a _____.

5. The force that initially causes the wind to blow is the _____ _____ _____.

6. At sea level, the average or standard value of atmospheric pressure is _____ millibars, _____ inches of mercury, and _____ hectopascals.

7. An elongated region of low pressure is called a _____.

8. An instrument that measures pressure but indicates altitude is the _____.

Multiple Choice

Top —
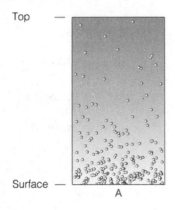

Surface —
A   B   C

FIGURE 1

1. In Figure 1, if the air temperature is the same in each column, which column has the highest surface air pressure? (Each dot represents billions of air molecules.)

   a. column A
   b. column B
   c. column C

2. The Coriolis force is strongest when the wind speed is _____ and the latitude is _____.

    a. high, low
    b. high, high
    c. low, high
    d. low, low

3. Another name for a constant pressure chart is:

    a. isobaric map
    b. zonal map
    c. meridional map
    d. gradient map
    e. geostrophic map

4. A barometer that contains no fluid:

    a. profiler
    b. anemometer
    c. aerovane
    d. aneroid

5. Wind flow pattern that generally moves from west to east is:

    a. meridional
    b. hydrostatic
    c. zonal

6. The surface winds around an area of low pressure normally _____. Above the system, the winds normally _____.

    a. diverge, converge
    b. diverge, diverge
    c. converge, diverge
    d. converge, converge

7. In the Northern Hemisphere directly above you (at about 10,000 ft), clouds are moving from south to north, indicating a south wind. From this observation you know that the center of lowest pressure aloft must be to the_____ of you.

    a. west
    b. east
    c. north
    d. south

8. On a weather map, the strongest winds are normally observed:

    a. at the center of high pressure
    b. at the center of low pressure
    c. near a large body of water
    d. where the isobars or contour lines are close together

9. A wind that blows at a constant speed parallel to *curved* lines above the level of friction is called a:

    a. geostrophic wind
    b. cyclonic wind
    c. gradient wind

True–False

_____ 1. The pressure gradient force is directed from high pressure toward lower pressure at all places on the earth.

_____ 2. To obtain *station pressure* with a mercury barometer, normally you must make a correction for altitude.

_____ 3. The Coriolis force causes the wind to blow faster.

_____ 4. The Coriolis force is the result of horizontal differences in pressure.

_____ 5. The winds aloft in middle latitudes of both hemispheres blow primarily from the west because the air aloft above high latitudes is colder than the air aloft above low latitudes.

_____ 6. If the earth stopped rotating, surface winds would blow directly from higher pressure toward lower pressure.

_____ 7. On an *upper level chart* (such as a 500-millibar chart), winds tend to cross the isobars or contours at an angle that averages about 30°.

_____ 8. Cold air aloft is usually associated with low pressure and warm air aloft with high pressure.

_____ 9. The Coriolis force is zero at the equator.

_____ 10. As the wind moves in a curved path, the centripetal force results from an imbalance between the Coriolis force and the pressure gradient force.

_____ 11. An instrument that indicates both wind speed and wind direction is the aerovane.

_____ 12. Generally, the top of the friction layer (planetary boundary layer) is about 3000 ft above the surface.

_____ 13. The air above a region of surface high pressure is normally rising.

_____ 14. During a south wind, a wind vane will point toward the north.

_____ 15. Atmospheric pressure decreases most rapidly in a cold column of air.

_____ 16. The air flow around an area of high pressure is called cyclonic flow.

_____ 17. On a weather map, closely spaced isobars or contour lines indicate a region of high winds because in this region there is a strong pressure gradient force.

_____   18.   An onshore wind blows from water onto the land.

_____   19.   An anticyclone is an area of low pressure.

Additional Questions

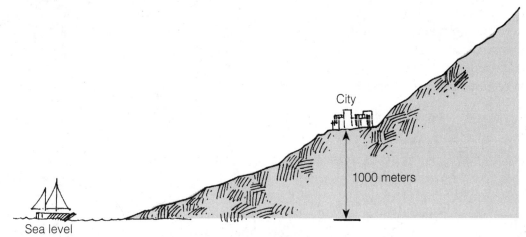

FIGURE 2

1.   What is the approximate sea level pressure for the city in Figure 2 when its station pressure is 920 millibars? (Use a pressure change of 10 millibars per 100 meters.)

_____ millibars

2.   Each of the four illustrations in Figure 3 represents the wind-flow pattern around an area of surface high or low pressure in either the Northern or Southern Hemisphere.

   a.   In the center of each pressure system indicate with the letter L or H whether it is a low or high pressure area.

   b.   In the space provided beneath each pressure system, write in whether this system is located in the Northern Hemisphere or Southern Hemisphere.

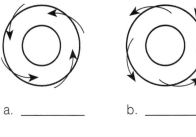

a. _____   b. _____

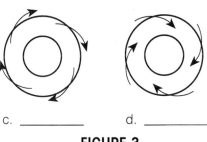

c. _____   d. _____

FIGURE 3

3. In Figure 4 show with arrows how the wind would blow around the area of low pressure and the area of high pressure on the upper-air chart. Both are located in the Northern Hemisphere and both are found aloft, above the level of friction. The lines around the L and H represent contour lines or isobars.

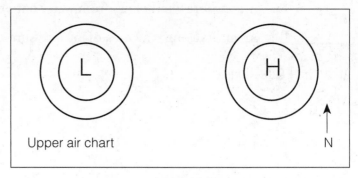

FIGURE 4

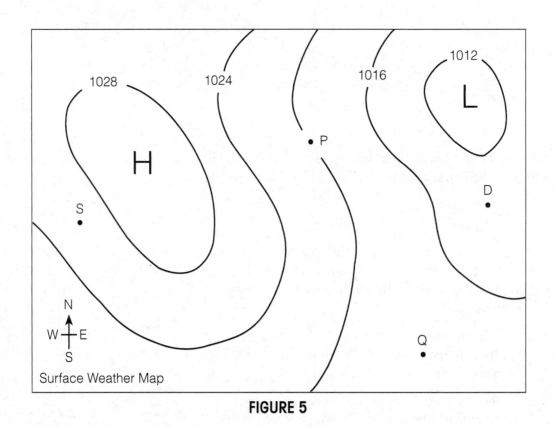

FIGURE 5

4. Answer the following questions that pertain to Figure 5, a surface weather map in the Northern Hemisphere.

    a. What is the sea level pressure at point P? _____

    b. At point P the wind would *most likely* be blowing from the _____.

    c. Would the pressure gradient force at point P be directed toward the H, the L, point D, or point Q? _____

    d. Would the wind at point S most likely be blowing from the southeast, southwest, northeast, or northwest? _____

e.   Would you expect the strongest winds at point P, point D, or point Q? _____
     Explain.

f.   Should the air above the L be rising or sinking? _____

     Should the air above the H be rising or sinking? _____

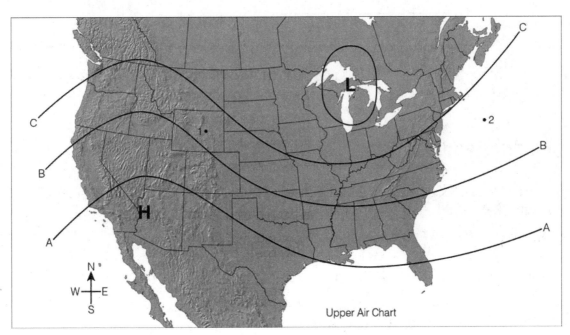

FIGURE 6

5.   Answer the following questions that pertain to Figure 6, a 500-millibar chart.

a.   The solid lines on the map are contour lines, indicating elevation above sea level. Which
     contour line (A, B, or C) represents the highest pressure? _____

b.   Between contour line A and contour line B, draw arrows to show the most likely wind-
     flow pattern for the map.

c.   Is the wind direction at point 1 most likely northwest, northeast, southwest, or south-
     east? _____ Is the wind direction at point 2 most likely northwest, north-
     east, southwest or southeast? _____

d.   Would stronger winds be observed at point 1 or point 2? _____ Explain.

6. Express the following wind directions in terms of compass points. (Hint: see Fig. 6.24, p. 161, in your textbook.)

Wind direction (degrees)	Wind direction (compass points)
270°	_____
360°	_____
90°	_____
225°	_____
315°	_____
135°	_____

Additional Readings

"Computer Map Analysis: Drawing Contours" by Alfred Blackadar, *Weatherwise*, Vol. 42, No. 2 (April 1989), p. 109.

"How Strong is the Wind?" by Frank H. Forrester, *Weatherwise*, Vol. 39, No. 3 (June 1986), p. 147.

"Secrets in the Wind" by David Bloom, *Weatherwise*, Vol. 51, No. 5 (September/October 1998), p. 14.

"Iceberg in the Sky" by Eric Pinder, *Weatherwise*, Vol. 51, No. 6 (November/December 1998), p. 16.

"The Big Wind" by Richard W. O'Donnell, *Weatherwise*, Vol. 51, No. 6 (November/December 1998), p. 20.

"The Great Coriolis Conspiracy" by Graham Kearns, *Weatherwise*, Vol. 51, No. 3 (May/June 1998), p. 36.

Answers

- -

Matching

1.	g	5.	i	9.	h	13.	m
2.	j	6.	c	10.	d	14.	f
3.	a	7.	n	11.	k	15.	e
4.	l	8.	o	12.	b		

Fill in the Blank

1. Coriolis force
2. Isaac Newton
3. contour
4. barograph
5. pressure gradient force

6. 1013.25 millibars; 29.92 inches; 1013.25 hPa
7. trough
8. altimeter

Multiple Choice

1.	b	4.	d	7.	a
2.	b	5.	c	8.	d
3.	a	6.	c	9.	c

True–False

1.	T	6.	T	11.	T	16.	F
2.	F	7.	F	12.	T	17.	T
3.	F	8.	T	13.	F	18.	T
4.	F	9.	T	14.	F	19.	F
5.	T	10.	T	15.	T		

Additional Questions

1. 1020 millibars
2. a. low pressure area, N.H.
    b. high pressure area, S.H.
    c. high pressure area, N.H.
    d. low pressure area, S.H.
3. Around the low, the wind blows counterclockwise and parallel to the lines. Around the high, the wind blows clockwise and parallel to the lines.

4. a. 1020 millibars
    b. northwest
    c. the L
    d. southeast
    e. point P; the isobars are closer together there, producing a stronger pressure gradient force.
    f. above the L, rising; above the H, sinking

5. a. Contour line A represents the highest altitude and also the highest pressure.

    b. The arrows should move from west to east *parallel* to the lines.

    c. at point 1, northwest; at point 2, southwest

    d. at point 1. The contour lines are closer together at point 1, which should produce stronger winds there.

6. 270°—west wind
    360°—north wind
    90°—east wind
    225°—southwest wind
    315°—northwest wind
    135°—southeast wind

Atmospheric Circulations

7

Chapter Seven focuses on a variety of atmospheric circulations. The chapter begins by introducing the hierarchy of atmospheric motions—what scientists call the scales of motion. This is followed by a section that describes how turbulent, twisting eddies form. The next several sections examine winds on a slightly larger scale. At this point we learn the causes of such winds as the sea breeze, the chinook, and the Santa Ana. The chapter then examines the large-scale circulations of air. Here we learn that there are semipermanent pressure systems that tend to persist in particular regions throughout the year. Winds that blow in response to these surface pressure systems establish prevailing wind patterns, such as the middle latitude westerlies, the polar easterlies, and the subtropical trade winds. The annual shifting of the pressure systems strongly influences the prevailing winds and annual precipitation patterns of many regions. After a section on jet streams, the chapter details the many interactions that take place between the atmosphere and the oceans.

Some important concepts and facts of this chapter:

1. A sudden change in wind speed or wind direction (or both) is called wind shear.

2. The sea breeze blows from the surface water onto the land in response to local pressure differences created by the uneven heating and cooling rates of land and water.

3. Monsoon winds are those that change direction seasonally. They usually blow from water onto land during the warm (wet) season, and from the land out over the water during the cool (dry) season.

4. Chinook winds are warm, dry winds that warm by compressional heating as they descend the leeward side of mountains.

5. The Santa Ana wind is a warm, dry wind that warms by compressional heating as it descends the high plateau into Southern California.

6. The subtropical high-pressure areas with their sinking air and clear skies are primarily responsible for the major deserts of the world located near 30° latitude.

7. The polar front is a zone of low pressure where storms often form. It separates the mild westerlies from the cold, polar easterlies.

8. The trade winds are located equatorward of the subtropical highs in both hemispheres.

9. Near the equator, the intertropical convergence zone (ITCZ) is a boundary where air rises in response to the flowing together of the northeast trades and the southeast trades.

10. In the Northern Hemisphere, the major global pressure systems and wind belts shift northward in summer and southward in winter.

11. Jet streams exist where strong winds become concentrated in narrow bands. They often form where sharp temperature changes produce rapid changes in pressure, such as aloft in the vicinity of the polar front.

12. Surface winds blowing over the ocean drive the major ocean currents. The currents, in turn, release energy to the atmosphere, which helps the atmosphere maintain its general circulation of winds.

13. A major El Niño event is a condition where warm surface water covers vast areas of the tropical Pacific. When the water in the equatorial Pacific turns colder than normal, this condition is called La Niña.

Self Tests

Match the Following

_____	1.	Wind blowing past a chimney is an example of this scale of motion
_____	2.	Sudden change in wind speed or wind direction
_____	3.	Warming of the surface water of the tropical Pacific west of South America
_____	4.	Abbreviation for clear air turbulence
_____	5.	Pressure systems that almost circle the globe near latitude 30°
_____	6.	The sea breeze is an example of this scale of motion
_____	7.	The rising of cold water
_____	8.	Boundary separating the northeast trades and the southeast trades
_____	9.	Violent eddies that form beneath the wave crest downwind of a mountain
_____	10.	The reversal of surface air pressure at opposite ends of the tropical Pacific Ocean
_____	11.	Semipermanent pressure system associated with the polar front
_____	12.	Warm ocean current that flows northward along the east coast of North America
_____	13.	The largest scale of atmospheric motions
_____	14.	The shallow, cold surface anticyclone observed over Asia during winter
_____	15.	The cold water episode that finds strong trade winds and unusually cold surface water in the central and eastern Pacific

a. rotors

b. subtropical highs

c. La Niña

d. subpolar low

e. microscale

f. Southern Oscillation

g. upwelling

h. El Niño

i. ITCZ

j. Siberian high

k. wind shear

l. Gulf stream

m. mesoscale

n. CAT

o. planetary scale

Additional Matching (Winds)

_____	1.	Wind system that changes direction seasonally
_____	2.	Wind belt observed behind the polar front
_____	3.	Another name for a whirlwind
_____	4.	A chinook wind in the Alps
_____	5.	A strong, often cold, downslope wind
_____	6.	Fast flowing current of air concentrated in a narrow band
_____	7.	The nighttime counterpart of the sea breeze
_____	8.	Cumulus clouds developing above isolated mountain peaks are often the result of this
_____	9.	Warm, dry wind associated with devastating fires in Southern California
_____	10.	A windstorm composed of dust and sand that forms along the leading edge of a thunderstorm
_____	11.	Wind belt found equatorward of the subtropical highs
_____	12.	Warm, dry wind that brings relief from the bitter cold on the eastern side of the Rockies

a.   Santa Ana

b.   foehn

c.   land breeze

d.   dust devil

e.   chinook

f.   polar easterlies

g.   jet stream

h.   monsoon

i.   trade winds

j.   valley breeze

k.   haboob

l.   katabatic wind

Fill in the Blank

1.   A cool, summertime breeze that blows from sea to land: _____ _____.

2.   Name of the warm, dry downslope wind observed on the eastern side of the Rocky Mountains: _____ _____.

3.   Circulations brought on by changes in air temperature are called _____ circulations.

4.   The majority of the United States lies within a wind belt known as the _____.

5.   _____ is mainly responsible for cold coastal water observed along the west coast of North America.

6.   Downslope winds warm by the process of _____ heating.

7.   _____ _____ are responsible for the existence of the major deserts of the world observed near 30° latitude.

8. The name given to the warm, dry downslope wind that blows from the east or northeast into Southern California: _____ _____ _____.

9. The jet stream that forms along the polar front is called the _____ _____ _____.

10. The Pacific Ocean's counterpart to the Atlantic's Bermuda high is called _____ _____.

Multiple Choice

1. In humid climates, clouds at night tend to form over the water during a:

    a. Santa Ana wind
    b. chinook wind
    c. lake breeze
    d. land breeze
    e. sea breeze

2. The winter monsoon in eastern and southern Asia is characterized by:

    a. wet weather and winds blowing from sea to land
    b. wet weather and winds blowing from land to sea
    c. dry weather and winds blowing from sea to land
    d. dry weather and winds blowing from land to sea

3. A thermal low does *not*:

    a. form in a region of warm surface air
    b. form in response to variations in surface air temperature
    c. become stronger with increasing height

4. In terms of the three-cell model of the general circulation, areas of surface low pressure should be found near:

    a. the equator and 60° latitude
    b. the equator and the poles
    c. the equator and 30° latitude
    d. 30° latitude and 60° latitude
    e. 30° latitude and the poles

5. Which of the winds listed below is blowing uphill?

    a. chinook wind
    b. mountain breeze
    c. katabatic wind
    d. Santa Ana wind
    e. valley breeze

6.   The horse latitudes are the result of:

   a.   the polar front jet stream
   b.   the ITCZ
   c.   the subtropical highs
   d.   the subpolar lows
   e.   the polar high

7.   According to the three-cell general circulation model, one would expect the driest regions of the world to be near:

   a.   latitude 30° and latitude 60°
   b.   latitude 30° and the polar regions
   c.   the equator and the polar regions
   d.   the equator and 30° latitude
   e.   the equator and 60° latitude

8.   Which of the following does *not* occur during an El Niño/Southern Oscillation (ENSO) event?

   a.   extensive ocean warming occurs over the tropical Pacific Ocean
   b.   it may last for many months
   c.   large numbers of fish often die along the coast of Peru
   d.   it influences the westerly winds aloft, bringing too much rain to some regions and too little to others
   e.   the subtropical high pressure area off the coast of South America increases in strength

9.   The main reason for the dry summers observed along the southwest coast of North America is that:

   a.   the air along the coast is too cool to produce adequate precipitation
   b.   the ITCZ moves north in summer, blocking storms from entering the area
   c.   the polar jet stream moves over the area in summer
   d.   the northeast trades sweep into the region
   e.   the Pacific high moves northward, producing sinking air over the region

10.   The three-cell model of the general circulation says that in the Northern Hemisphere, you would expect to observe the westerlies:

   a.   southward of the ITCZ
   b.   southward of the subtropical highs
   c.   northward of the subtropical highs
   d.   northward of the subpolar lows
   e.   southward of the northeast trades

11. At the equator, according to the three-cell general circulation model, we would *not* expect to find:

    a. heavy showers
    b. the ITCZ
    c. cumuliform clouds
    d. a ridge of high pressure

12. In the general circulation of the atmosphere, the doldrums are found:

    a. over equatorial waters
    b. along the polar front
    c. in the center of the subtropical highs
    d. in the Icelandic low
    e. near the polar high

True–False

_____   1. Along the coast in summer, a sea breeze is usually strongest and best developed in the afternoon.

_____   2. Dust devils usually descend from the base of a cumulonimbus cloud.

_____   3. Lines of equal wind speed are called isotachs.

_____   4. The California current is a warm ocean current that flows parallel to the west coast of North America.

_____   5. In the Northern Hemisphere, the major features of the general circulation shift northward in summer and southward in winter.

_____   6. The region of strong winds of a jet stream is called a jet maximum or jet streak.

_____   7. Chinook wall clouds often signal the possible onset of a chinook.

_____   8. The polar jet stream is strongest in winter and moves farther south in summer.

_____   9. Average winter temperatures in Great Britain and Norway would probably be much colder if it were not for the Canary current.

_____   10. The polar jet tends to flow in a wavy east-to-west pattern.

_____   11. Clear air turbulence can occur in the vicinity of a jet stream.

_____   12. The polar front separates the cold, polar easterlies from the mild westerlies.

_____   13. The summer monsoon in eastern and southern Asia is characterized by wet weather and winds that blow from sea to land.

_____   14. In the Northern Hemisphere, the polar front jet stream actually separates colder air to the north and relatively mild air to the south.

_____   15. The subtropical jet stream is normally found equatorward of the polar jet stream.

_____   16.   The reversal of surface water temperatures in the north Pacific Ocean that occurs every 20 to 30 years is called the Pacific Decadal Oscillation.

_____   17.   Cold relatively dry winters in Northern Europe and along the east coast of North America are generally associated with the negative phase of the North Atlantic oscillation.

Additional Questions

FIGURE 1

1.   Figure 1 represents summertime conditions along the coast of a large body of water during the afternoon.

    a.   In the diagram, label where the surface air is relatively warm and where it is relatively cool.

    b.   Place an H in the diagram where the surface air pressure would be relatively high and an L where the surface air pressure would be relatively low.

    c.   Draw several arrows to show the direction of the surface wind.

    d.   The surface wind you have drawn in (c) would be called a _____ _____.

    e.   Draw several more arrows and complete the thermal circulation.

    f.   Draw a cloud to show where you would expect one to form. Explain why you chose that location.

2.   In the circle representing the earth (Figure 2), place the major semipermanent surface pressure systems at their appropriate latitudes. Label them properly and be sure to include the polar front and the ITCZ. (Hint: Use the three-cell model of the general circulation, Fig. 7.21, p. 185, in your textbook.)

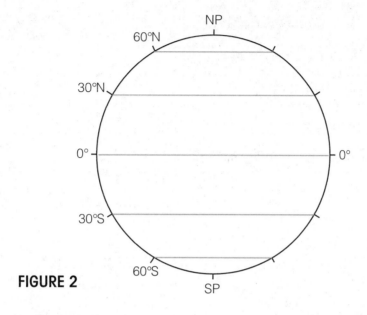

FIGURE 2

3.   In the circle representing the earth (Figure 3) draw arrows to represent the major surface wind belts of the world. Label them correctly. (Hint: Again use the three-cell model, Fig. 7.21 in your textbook.)

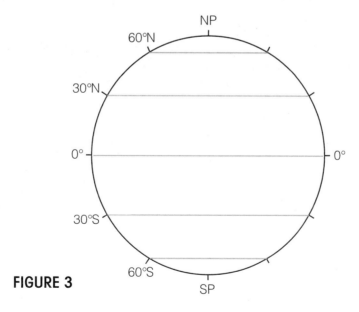

FIGURE 3

4. Use Figures 2 and 3 and the three-cell model of the general circulation of the atmosphere, Fig. 7.21, p. 185, in your textbook, to predict the prevailing winds for each of the following cities:

Prevailing Winds

a. Chicago, Illinois (latitude 42°N)

_____

b. Barrow, Alaska (latitude 70°N)

_____

c. Winnipeg, Canada (latitude 50°N)

_____

d. Miami, Florida (latitude 26°N)

_____

e. Auckland, New Zealand (latitude 37½°S)

_____

f. Launda, Angola (latitude 9°S)

_____

Additional Readings

"Monsoons" by Priit J. Vesilind, *National Geographic*, Vol. 166, No. 6 (December 1984), p. 712.

"The Santa Ana Wind of Southern California" by Arthur G. Lessard, *Weatherwise*, Vol. 41, No. 2 (April 1988), p. 100.

"Utah's Great Salt Lake—A Classic Lake Effect Snowstorm" by David M. Carpenter, *Weatherwise*, Vol. 38, No. 6 (December 1985), p. 309.

"Dust Devils in the Desert" by Andy Woodcock, *Weatherwise*, Vol. 44, No. 4 (Aug–Sept 1991), p. 39.

"Chinook Winds Resemble Water Flowing over a Rock" by Richard A. Kerr, *Science*, Vol. 231, No. 4743 (March 14, 1986), p. 1244.

"El Niño and La Niña" by George Philander, *American Scientist*, Vol. 77, No. 5 (Sept–Oct 1989), p. 451.

"Written in the Winds" by Jerome Namias, *Weatherwise*, Vol. 42, No. 2 (April 1989), p. 85.

"All About the Jet Stream" by Thomas Schlatter, *Weatherwise*, Vol. 40, No. 1 (February 1987), p. 50.

"Upslope, Downslope" by Mark Hanak, *Weatherwise*, Vol. 47, No. 6 (December 1994/January 1995), p. 28.

"California Crazy" by Dan Graf, *Weatherwise*, Vol. 48, No. 6 (December 1995/January 1996), p. 28.

"Clearing the Air about Turbulence" by Ron Cowen, *Weatherwise*, Vol. 51, No. 6 (November/December 1998), p. 24.

"Billow Talk" by Lee Grenci, *Weatherwise*, Vol. 52, No. 6 (November/December 1999), p. 33.

"Winds of the World'" by Jan Null, *Weatherwise*, Vol. 53, No. 3 (May/June 2000), p. 36.

"The El Niño Factor" by Carl Zimmer, *Discover*, Vol. 20, No. 1 (January 1999), p. 98.

"The Long Haul" by Stephanie Ocko, *Weatherwise*, Vol. 50, No. 6 (December 1997), p. 12.

"Close Encounter with a Rocky Mountain Whirlwind" by Dennis McCown, *Weatherwise*, Vol. 50, No. 3 (June/July 1997), p. 29.

"Underwater Weather," Robert Monroe, *Weatherwise*, Vol. 55, No. 3 (May/June 2002), p. 15.

"Wild Winds of the Andes," Ed Darack, *Weatherwise*, Vol. 55, No. 4 (July/August 2002), p. 28.

"The Moving Rocks of Death Valley," Ed Darack, *Weatherwise*, Vol. 56, No. 1 (January/February 2003), p. 18.

"Atlantic Rhythms," John Weier, *Weatherwise*, Vol. 56, No. 3 (May/June 2003), p. 28.

"Monsoon Madness," by Roff Smith, *National Geographic*, Vol. 206, No. 5 (November 2004), p. 86.

"Tiny Tornadoes," by Kathy A. Smith, *Discover*, Vol. 24, No. 7 (July 2003), p. 44.

"Sky Power," by Ed Darack, *Weatherwise*, Vol. 58, No. 6 (November/December 2005), p. 30.

"The Wellington Winds," by Lina Beeman, *Weatherwise*, Vol. 57, No. 6 (November/December 2004), p. 42.

"El Niño Forever," by David Laskin, *Weatherwise*, Vol. 58., No. 6 (November/December 2005), p. 48.

"A Break in the Clouds," by Sarah Sealia, *Weatherwise*, Vol. 59, No. 1 (January/February 2006), p. 42.

Answers

Matching

1.	e	5.	b	9.	a	13.	o
2.	k	6.	m	10.	f	14.	j
3.	h	7.	g	11.	d	15.	c
4.	n	8.	i	12.	l		

Additional Matching (Winds)

1.	h	4.	b	7.	c	10.	k
2.	f	5.	l	8.	j	11.	i
3.	d	6.	g	9.	a	12.	e

Fill in the Blank

1. sea breeze
2. chinook wind
3. thermal
4. westerlies
5. upwelling
6. compressional
7. subtropical highs or subtropical anticyclones
8. Santa Ana wind
9. polar front jet stream or polar front jet
10. Pacific high or Pacific anticyclone

Multiple Choice

1.	d	4.	a	7.	b	10.	c
2.	d	5.	e	8.	e	11.	d
3.	c	6.	c	9.	e	12.	a

True–False

1.	T	6.	T	11.	T	16.	T
2.	F	7.	T	12.	T	17.	T
3.	T	8.	F	13.	T		
4.	F	9.	F	14.	T		
5.	T	10.	F	15.	T		

Additional Questions

1. a. The relatively warm air is above the land; the relatively cool air is above the water.
    b. The surface H should appear above the water; the surface L above the land.
    c. The surface wind will blow from the water onto the land.
    d. Sea breeze
    e. To complete the circulation the air should be rising over the land. Aloft, the air should be moving from land to water, with air sinking over the water.
    f. The cloud should be drawn over the land where the air is rising, expanding and cooling.

2. and 3.   Answers are found in Fig. 7.21,
        p. 185, in your textbook.
4.   Predicted prevailing winds:
    a.   westerly
    b.   easterly or northeasterly
    c.   westerly
    d.   northeasterly
    e.   westerly
    f.   southeasterly

Air Masses, Fronts, and Middle Latitude Cyclones

8

The first part of Chapter Eight examines the various air masses and the weather each brings to a particular region. Here we learn that an air mass has fairly uniform properties of temperature and moisture in any horizontal direction, and for an air mass to acquire such properties it must originate over generally flat terrain of uniform composition, where the winds are fairly light. Where air masses with sharply contrasting properties meet, we find weather fronts. Consequently, the next several sections examine the structure, weather, and development of fronts. The latter part of the chapter describes where, why, and how a middle latitude cyclonic storm forms and moves.

Some important concepts and facts of this chapter:

1. An air mass is a large body of air with similar horizontal temperature and humidity characteristics.

2. Regions where air masses originate and acquire their properties of temperature and moisture are called source regions. These source regions are generally flat, of uniform composition, with light winds.

3. Continental (c) air masses form over land. Maritime (m) air masses form over water. Polar (P) air masses originate in cold, polar latitudes and extremely cold air masses are designated as arctic (A). Tropical (T) air masses originate in warm, tropical latitudes.

4. cP air masses are cold and dry; cA air masses are extremely cold and dry; cT air masses are hot and dry; mT air masses are warm and moist; mP air masses are cold and moist.

5. A front is a transition zone that separates two air masses with contrasting properties, usually temperature and humidity.

6. A cold front is a region where colder air is replacing warmer air. Typically, along the leading edge of the cold front, warmer air is forced upward producing a rather narrow band of showers.

7. A warm front is a region where warmer air is replacing colder air. Along the warm front, warmer air rides up and over colder surface air, producing clouds and precipitation out ahead of the advancing surface front.

8. An occluded front often has characteristics of both a cold front and a warm front. The coldest air is observed *behind* a cold occlusion and *ahead* of a warm occlusion.

9. Conceived in Norway, the polar front theory is a model of how an ideal storm progresses through the stages of birth, maturity, and finally, dissipation.

10. For a surface mid-latitude cyclonic storm to develop or intensify, the upper-level low must be located to the west of (behind) the surface low.

11. For a surface mid-latitude cyclonic storm to develop or intensify, upper-level divergence of air above the storm must be greater than surface convergence of air (more air must be removed at the top than is brought in at the surface).

12. As the polar front jet stream sweeps slightly south of a developing surface storm, it provides some of the necessary ingredients for the storm's development.

Self Tests

. . . .

Match the Following

_____   1.   This front is drawn in red on a weather map

_____   2.   The development or strengthening of a mid-latitude cyclone

_____   3.   The piling up of air above a region

_____   4.   The rising of warm air over cold air

_____   5.   Another name for a large, mid-latitude cyclonic storm

_____   6.   This front is drawn in purple on a weather map

_____   7.   The spreading out of air above a region

_____   8.   Storm systems that tend to form on the downwind side of a mountain range

_____   9.   Heavy snow showers that fall on the eastern shores of large lakes

_____   10.   This front is drawn in blue on a weather map

_____   11.   Small disturbance imbedded in a longwave

_____   12.   A kink that forms along a front

_____   13.   Another name for Rossby waves

_____   14.   An extremely large body of air whose properties of temperature and moisture are similar in any horizontal direction

_____   15.   Storms that form along the eastern seaboard of the United States, then move northeastward along the coast

a.   overrunning

b.   shortwave

c.   warm front

d.   leeside lows

e.   frontal wave

f.   convergence

g.   northeasters

h.   occluded front

i.   cyclogenesis

j.   lake-effect snows

k.   wave cyclone

l.   longwaves

m.   divergence

n.   cold front

o.   air mass

Additional Matching (Air Masses)

(Some answers may be used more than once.)

_____ 1. Air mass responsible for hot, muggy weather in the eastern half of North America

_____ 2. The coldest of all air masses

_____ 3. An air mass that forms over North America only in summer

_____ 4. Air mass responsible for refreshing cool, dry breezes after a long, humid summer hot spell in the Central Plains

_____ 5. Air mass responsible for heavy rain, flooding, mudslides and the melting of snow at high elevations in Southern California

_____ 6. Air mass responsible for hot, dry summer weather over portions of the arid southwestern United States

_____ 7. During the winter this air mass is responsible for heavy snow showers on the western slopes of the Sierra Nevada and Cascade Mountains

_____ 8. Air mass responsible for daily afternoon thunderstorms along the Gulf Coast states

_____ 9. Record-breaking low temperatures during the middle of winter are most often associated with this air mass

_____ 10. Cold, moist air mass that moves into northeastern North America from off the Atlantic Ocean

_____ 11. Chinook winds often develop when this air mass replaces very cold continental polar air

a. cP

b. cT

c. cA

d. mP

e. mT

Fill in the Blank

1. In the space next to each air mass letter designation, write in what the letters represent. Also describe the temperature and moisture characteristics of the air mass.

Letter Designation	Name	Temperature/Moisture Characteristics
cP		
cT		
cA		
mP		
mT		

2. For a surface mid-latitude cyclone to develop or intensify into a deep low pressure area, aloft _____ must be greater than the surface convergence of air.

3. On a surface weather map, the transition zone between two air masses with sharply contrasting properties is called a _____.

4. On a surface weather map a _____ front represents a region where colder air is replacing warmer air.

5. On a surface weather map where cold air is replacing cool air, an _____ front is drawn.

6. On a surface weather map a _____ front represents a region where warmer air is replacing colder air.

7. As the _____ _____ bends into a looping wave pattern, it provides some of the necessary ingredients for a developing surface storm system.

8. For a surface mid-latitude cyclone to develop or intensify, the upper level low must be to the _____ of the surface low.

Multiple Choice

1. The origin of cA and cP air masses that enter the United States is:

    a. the North Atlantic Ocean
    b. northern Canada and Alaska
    c. northern Siberia
    d. the North Pacific Ocean
    e. the desert southwest

2. When an upper-level low lies directly above a midlatitude storm system, the surface low will usually:

    a. dissipate
    b. intensify
    c. show no change during a 48-hour period

3. During the winter, cold, dry air will occasionally move into Washington, Oregon, and California from the east and northeast. However, by the time this cold air reaches the coastal regions it is often much warmer than it was originally primarily because:

    a. the air sinks, compresses and warms
    b. friction with the ground warms the air
    c. the sun heats the air
    d. the ocean warms the air
    e. latent heat of condensation warms the air as it moves downhill

4. The greatest contrast in both *temperature* and *moisture* will occur along the boundary separating which air masses?

    a. cP and cT in summer
    b. cP and mT in summer
    c. mP and mT in summer
    d. mP and mT in winter
    e. cA and mT in winter

5. Which below is *not* a name given to a large cyclonic storm system that forms in the middle latitudes?

    a. wave cyclone
    b. middle latitude cyclone
    c. anticyclone
    d. extratropical cyclone

6. According to the polar front theory of a developing wave cyclone, the storm system is usually most intense:

    a. as a stable wave
    b. as a stationary front
    c. as an open wave
    d. as a frontal wave
    e. when the system first becomes occluded

7. What type of weather front would be responsible for the following weather forecast: "Increasing high cloudiness and cold this morning. Clouds increasing and lowering this afternoon with a chance of snow or rain tonight. Precipitation ending by noon tomorrow. Turning much warmer. Winds light easterly today becoming southeasterly tonight and southwesterly by tomorrow."

    a. warm front
    b. cold front
    c. warm-type occluded front
    d. cold-type occluded front

8. What type of weather front would be responsible for the following weather forecast: "Increasing cloudiness and warm today with the possibility of showers and thunderstorms by this evening. Turning much colder tonight. Winds southwesterly today, becoming gusty and shifting to northwesterly by tonight."

    a. cold-type occluded front
    b. warm-type occluded front
    c. warm front
    d. cold front

9. Typically, winter mP air masses along the Atlantic coast of North America are less common than mP air masses along the Pacific coast mainly due to the fact that:

    a. the water is warmer along the Atlantic coast
    b. the prevailing winds aloft are westerly
    c. the landmass along the Atlantic coast is colder
    d. the source region for mP air on the Atlantic coast is western Europe
    e. the water is much colder along the Pacific coast

10. The coldest air mass would be found behind which of the following fronts?

    a. arctic front
    b. polar front
    c. cold-occluded front
    d. warm-occluded front
    e. warm front

11. During the winter the upper air flow on the adjacent map would bring _____ air masses into western North America and _____ air masses into eastern North America.

    a. mP, mT
    b. cA, mP
    c. cA, mT
    d. cP, mP
    e. cA, cT

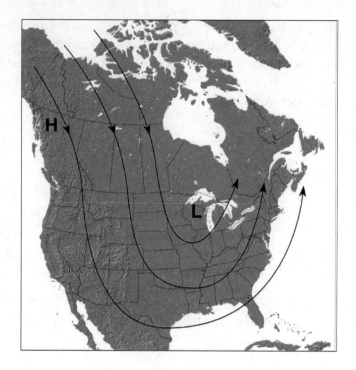

True–False

_____ 1. On a surface weather map, a cold front represents a trough of low pressure.

_____ 2. Generally, warm fronts move faster than cold fronts.

_____ 3. The rising of warm air and the sinking of cold air provides energy for a developing wave cyclone.

_____ 4. Wintertime thunderstorms are most often associated with cold fronts.

_____ 5. A good source region for an air mass is generally a windy region.

_____ 6. When a mid-latitude cyclonic storm is said to be *deepening* or *intensifying*, the air pressure at the center of the storm is dropping.

_____ 7. The polar front theory was conceived in the United States.

_____ 8. Occluded fronts may form as a cold front overtakes a warm front.

_____ 9. On a surface weather map, a stationary front is drawn in purple.

_____ 10. Persistent, hot, humid weather that may last for days in the southeastern United States could be classified as air mass weather.

_____ 11. Storms generally move faster in summer than in winter.

_____ 12. When a high pressure area (anticyclone) is said to be *building*, the air pressure at the center of the high is rising.

_____ 13. When an upper-level trough is located to the west of a surface mid-latitude storm, the surface storm will normally move toward the northeast.

_____ 14. Before the passage of a cold front the air pressure normally drops; after the passage, the air pressure normally rises.

_____ 15. At the surface, during the passage of a cold-type occluded front, the coldest air is observed ahead of the advancing front.

_____ 16. The slope of a typical warm front is usually much steeper than that of a cold front.

_____ 17. Shortwaves in the flow aloft tend to move quickly around longwaves.

_____ 18. Drylines usually separate cold moist air from extremely cold dry air.

Additional Questions

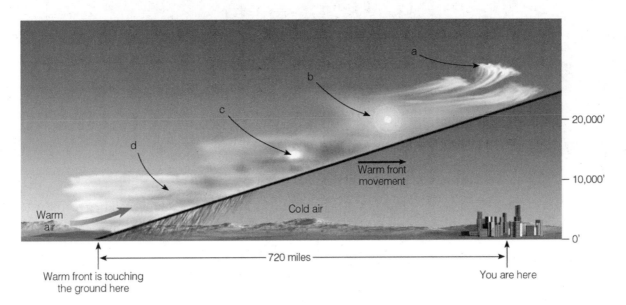

FIGURE 1

1. Figure 1 shows a warm front with precipitation moving toward your home in winter.

   a. Letters a through d represent a sequence of clouds that you might expect to observe as the warm front approaches you. List the clouds in their proper order below.

   cloud a _____

   cloud b _____

   cloud c _____

   cloud d _____

   b. Suppose that 720 miles from your home the warm front is at the surface. If the front is moving toward you at an average speed of 10 miles per hour, how long will it take before the front passes your area?

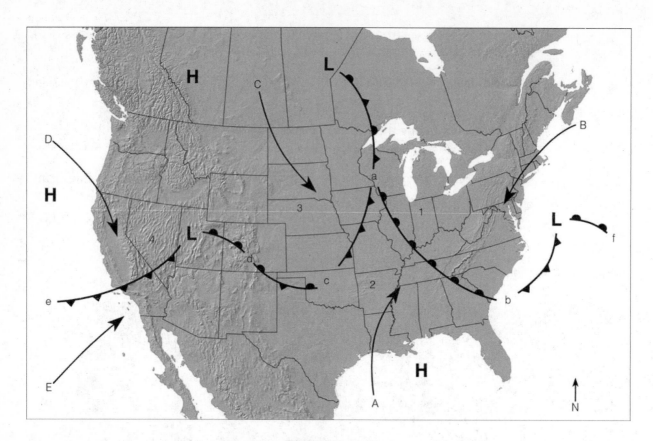

FIGURE 2

2.   Figure 2 is a surface weather map. In the space below, label the fronts that fall between the letters on the map.

Front between letters   **Name of front**

L and a

_____

L and b

_____

L and f

_____

L and e

_____

L and d

_____

a and b

_____

a and c

_____

c and d

_____

3. Figure 2 represents a winter weather map with the large arrows showing the movement of air masses. The letters below correspond to the letters on the map. Next to each letter place the type of air mass that is present on the map and its temperature and moisture characteristics.

Air Mass   **Characteristics**

A

_____

B

_____

C

_____

D

_____

E

_____

4. Look closely at Figure 2 and match the weather conditions listed below with the number that appears on the map.

a. Overcast, cold, sleet, southeasterly winds, falling air pressure. Number _____

b. Partly cloudy, windy, very cold, northwesterly winds, rising air pressure. Number _____

c. Partly cloudy, mild, southwesterly winds, falling air pressure. Number _____

d. Rainshowers, cool, northwesterly winds, rising air pressure. Number _____

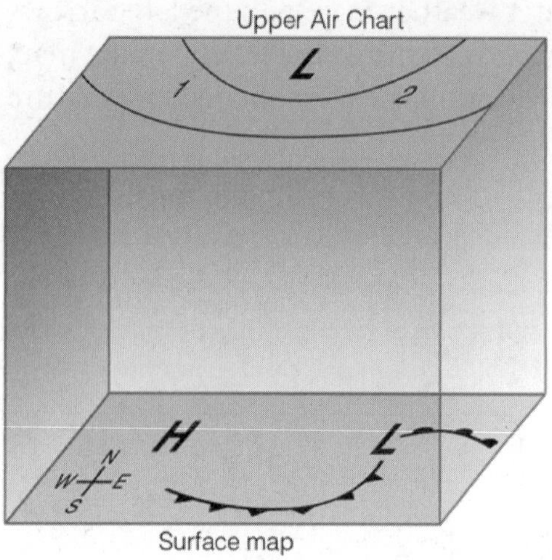

Upper Air Chart

Surface map

FIGURE 3

5.   Figure 3 shows a surface map and an upper-level chart.

   a.   On the upper-level chart:

   Divergence of air is probably occurring at point _____.

   Convergence of air is probably occurring at point _____.

   The wind at point 1 is probably blowing from the _____. (give direction)

   The wind at point 2 is probably blowing from the _____. (give direction)

   b.   On the surface weather map:

   According to the winds on the upper-level chart at point 2, the surface low pressure area should move toward the _____. (give direction)

   The storm system is in a stage of development called the _____ _____. (Hint: look at Fig. 8.23, p. 223, in your text.)

   Rising air would be observed above the high or low pressure area? _____

Additional Readings

"A Visible Cold Front" by Lynette Rummel, *Weatherwise*, Vol. 40, No. 4 (July–Aug 1987), p. 183.

"Bombs and Ultrabombs" by Bud Dorr, *Weatherwise*, Vol. 43, No. 2 (April 1990), p. 76.

"Arctic Hurricanes" by Steven Businger, *American Scientist*, Vol. 79, No. 1 (Jan–Feb 1991), p. 18.

"The Blizzard of '88'" by Patrick Hughes, *Weatherwise*, Vol. 40, No. 6 (December 1987), p. 312.

Skywatch: The Western Weather Guide by Richard A. Keen; Fulcrum, Inc., Golden, Colorado, 1987. See especially pp. 12–22.

Storms by William R. Cotton; ASTeR Press, Fort Collins, Colorado, 1990.

"The Blizzard of '93'" by Hank Brandli, *Weatherwise*, Vol. 46, No. 3 (June/July 1993), p. 9.

"Tragedy in Chicago" by Patrick Hughes and Douglas LeConte, *Weatherwise*, Vol. 49, No. 1 (February/March 1996), p. 18.

"Hurricanes in Disguise" by Robert Henson, *Weatherwise*, Vol. 48, No. 6 (December 1995/January 1996), p. 12.

"Chaos Rules" by Stanley David Gedzelman, *Weatherwise*, Vol. 47, No. 4 (August/September 1994), p. 21.

"New Respect for Nor'easters" by Ben Watson, *Weatherwise*, Vol. 46, No. 6 (December 1993/January 1994), p. 18.

"The Witch of November" by Mace Bentley and Steve Horstmeyer, *Weatherwise*, Vol. 51, No. 6 (November/December 1998), p. 29.

"Winter Storm Nicknames" by Mace Bentley, *Weatherwise*, Vol. 50, No. 6 (December 1997), p. 27.

"Name That Storm," by Sean Potter, *Weatherwise*, Vol. 58, No. 1 (January/February 2005), p. 24.

Answers

Matching

1.	c	5.	k	9.	j	13.	l
2.	i	6.	h	10.	n	14.	o
3.	f	7.	m	11.	b	15.	g
4.	a	8.	d	12.	e		

Additional Matching

1.	mT	4.	cP	7.	mP	10.	mP
2.	cA	5.	mT	8.	mT	11.	mP
3.	cT	6.	cT	9.	cA		

Fill in the Blank

1. cP, continental polar, cold, dry
    cT, continental tropical, hot, dry
    cA, continental arctic, very cold, dry
    mP, maritime polar, cold, moist
    mT, maritime tropical, warm, moist
2. divergence (of air)

3. front
4. cold
5. occluded
6. warm
7. jet stream or polar front jet stream
8. west (or left)

Multiple Choice

1.	b	4.	e	7.	a	10.	a
2.	a	5.	c	8.	d	11.	c
3.	a	6.	e	9.	b		

True–False

1.	T	6.	T	11.	F	16.	F
2.	F	7.	F	12.	T	17.	T
3.	T	8.	T	13.	T	18.	F
4.	T	9.	F	14.	T		
5.	F	10.	T	15.	F		

Additional Questions

1. a. cloud a, cirrus
        cloud b, cirrostratus
        cloud c, altostratus
        cloud d, nimbostratus
    b. it will pass in about 72 hours or 3 days
2. L and a, occluded front

L and b, cold front
L and f, warm front
L and e, cold front
L and d, warm front
a and b, warm front
a and c, cold front
c and d, stationary front

3. A. maritime tropical, mT, warm and humid
    B. maritime polar, mP, cold and moist
    C. continental Arctic (or polar), cA or cP, cold and dry
    D. maritime polar, mP, cold and moist
    E. maritime tropical, mT, warm and humid

4. a. Number 1
    b. Number 3
    c. Number 2
    d. Number 4

5. a. divergence of air at point 2
        convergence of air at point 1
        wind at point 1 is probably blowing from the NW
        wind at point 2 is blowing from the SW or WSW
    b. surface low should move toward the NE
        it is in the open wave stage
        rising air above the low pressure area

Weather Forecasting

9

Chapter Nine concentrates on predicting the weather. The beginning of the chapter describes some of the methods and procedures used in making a weather forecast. Here we learn that many ingredients—the current weather map, upper-air charts, satellite data, intuition, experience, and guidance from many computer progs—go into making a weather forecast. Woven into this section are some of the problems that confront anyone who attempts to predict the weather. Some of the modern techniques designed to improve weather forecasts are covered next. The latter part of the chapter considers how short-range weather forecasts can be made by observing the weather and by watching the movement of weather systems on a surface weather map.

Some important concepts and facts of this chapter:

1. A weather *watch* indicates that atmospheric conditions are favorable for the development of hazardous weather. A weather *warning* indicates that hazardous weather is either imminent or actually occurring.

2. A persistence forecast is a prediction that future weather will be the same as the present weather. If it is presently raining, a persistence forecast will call for rain.

3. A steady-state forecast (trend method) is a weather prediction based on the assumption that weather systems will continue to move in the same direction and at approximately the same speed as they have been moving.

4. The analogue method of forecasting makes a weather prediction by comparing past weather maps and weather patterns to those of the present. It is what forecasters call "pattern recognition."

5. Weather type forecasting categorizes weather patterns into similar groups or "types," using such criteria as the position of the subtropical highs and the upper-level flow.

6. Climatological forecasts are based on the climatology (average weather) of a particular region.

7. Ensemble forecasting is a technique based on running several forecast models (or different versions of a single model), each beginning with slightly different weather information to reflect errors in the measurements. If the different versions agree fairly well, a forecaster can place a high degree of confidence in the forecast. A low degree of confidence means that the models do not agree.

8. For a forecast to show skill, it must be better than a persistence forecast or a climatological forecast.

9. After a number of days, the atmosphere's chaotic behavior along with flaws in the computer models and small errors in the data greatly limit the accuracy of weather forecasts.

Self Tests

Match the Following

_____	1.	Forecast chart that shows the atmosphere at some future time
_____	2.	Satellites that remain above a fixed spot on the earth's surface
_____	3.	These use mathematical equations to describe the atmosphere's behavior
_____	4.	A surface or upper-level chart that shows various weather elements and patterns
_____	5.	This indicates that atmospheric conditions are favorable for the development of hazardous weather
_____	6.	Satellites that move in such a way as to closely parallel the earth's meridian lines
_____	7.	Forecasting method that assumes that weather systems will move in the same direction and at the same speed as they have been moving
_____	8.	This indicates that hazardous weather is imminent or actually occurring
_____	9.	Forecasting method that compares past weather maps and weather patterns to those of the present
_____	10.	Communication system that provides various weather maps and overlays on computer screens
_____	11.	An interaction between weather changes that occur in different regions of the world
_____	12.	Small unpredictable atmospheric fluctuations that produce uncertainties in a weather forecast
_____	13.	Chart that shows how weather variables have changed or will change over a given period of time.

a. atmospheric models

b. AWIPS

c. teleconnections

d. steady state/trend method

e. geostationary

f. analogue method

g. meteogram

h. prog

i. polar orbiting

j. watch

k. analysis

l. warning

m. chaos

Fill in the Blank

1. The forecasting of weather by high-speed computers is known as _____ _____ _____.

2. A weather forecast that is based on the weather conditions averaged over many years is called a _____ forecast.

3. If severe weather is imminent, a weather _____ is issued by the National Weather Service.

4. A weather forecast that predicts that the future weather will be the same as the present weather is called a _____ forecast.

Multiple Choice

1. A forecast that calls for a "chance of rain" suggests that the probability of rain is:

   a. 20 percent
   b. 30–50 percent
   c. 60–70 percent
   d. ≥ 80 percent

2. Suppose that where you live the weather during July is hot and humid, with afternoon thunderstorms. If for next July 4 you forecast the weather to be "hot and humid with afternoon thunderstorms," you would have made:

   a. a persistence forecast
   b. an ensemble forecast
   c. a climatological forecast
   d. a forecast based on the analogue method
   e. a forecast based on weather types

3. Suppose it is raining and cold outside. You look at the present 500-millibar chart and remember a similar weather pattern years ago where the cold rain actually changed into snow. If you use this information to make a forecast of "rain changing to snow," you would have employed which of the following forecasting methods?

   a. steady state/trend
   b. climatological
   c. ensemble
   d. analogue (pattern recognition)
   e. persistence

4. During changing weather patterns, the least accurate forecast method of predicting the weather two days into the future would be the method called:

    a. persistence
    b. analogue
    c. weather types
    d. numerical weather prediction

5. If a forecaster gives a "degree of confidence" with his or her weather forecast, then the forecaster used a forecasting technique called:

    a. steady state or trend
    b. climatology
    c. persistence
    d. analogue
    e. ensemble

6. Atmospheric models:

    a. are physical models that draw a picture of a developing storm system
    b. are copies of progs that were correctly used in forecasting the weather
    c. use mathematical equations to describe the behavior of the atmosphere
    d. describe atmospheric conditions using clay, blocks, and chart paper
    e. show off the latest attire in meteorological fashions

7. Which below is presently a problem with modern-day weather predictions?

    a. there are regions of the world where only sparse observations are available
    b. computer models do not always adequately interpret the surface's influence on the weather
    c. computer-forecast models make assumptions about the atmosphere that are not always correct
    d. the distance between grid points on some models is too large
    e. all of the above

8. If a cloud appears white in a visible satellite picture and gray in an infrared picture, then the cloud would most likely be a:

    a. low cloud
    b. high cloud
    c. middle cloud

9. If you live in a nonmountainous area, and the National Weather Service issues a forecast of 8 inches of snow over the next twenty-four hours, then this forecast will probably be accompanied by a:

    a. snow advisory
    b. winter storm warning
    c. blizzard warning

10. This type of weather forecast extends from about 3 to 8.5 days into the future and is based on computer-derived information:

    a. nowcast
    b. short-range forecast
    c. medium-range forecast
    d. long-range forecast

11. Which forecasting method is most closely associated with "pattern recognition"?

    a. persistence forecast
    b. analogue method
    c. climatological forecast
    d. steady state/trend
    e. ensemble

True–False

_____ 1. For a weather forecast to show skill it must be better than one based on persistence or climatology.

_____ 2. A probability forecast that calls for a 60 percent chance of rain means that it will rain 60 percent of the time over the forecast area.

_____ 3. High cirrus clouds will appear white in both a visible satellite picture and an infrared satellite picture.

_____ 4. An accurate forecast must show skill.

_____ 5. By examining a surface weather map, the movement of a surface low pressure area can be predicted based upon the orientation of the isobars in the warm air ahead of the cold front.

_____ 6. Infrared satellite pictures are computer enhanced to increase the contrast between specific features in the picture.

_____ 7. By examining a 500-millibar chart, you can determine the direction of movement of a surface low pressure area.

_____ 8. The ASOS system provides weather information about wind, temperature, pressure, and cloud height.

Additional Questions

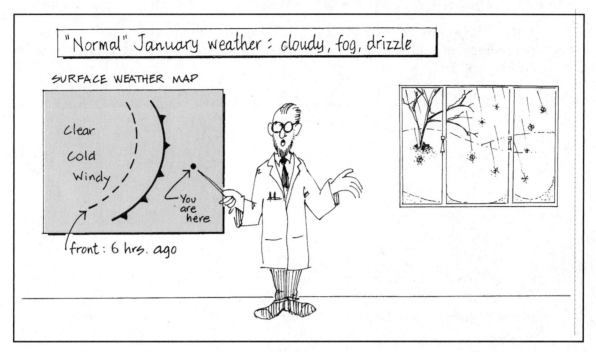

FIGURE 1

1. Suppose it is the end of January and the average weather for your area this time of year during the past 30 years or so has been overcast with fog and drizzle. But, as you can see in Figure 1, outside it is presently cold and snowing. However, a surface weather map indicates that a cold front is approaching from the west and will move through your area in about six hours. On the map you notice that directly behind the front the weather clears and turns much colder. With all of this information, you use the following methods to make a 12-hour weather forecast for your area.

    a.   A *persistence forecast* would be:

    b.   A *climatological forecast* would be:

    c.   A 12-hour weather forecast using the *steady state or trend method* would be:

2. Given the following weather conditions, make a short-range local weather forecast. (Hint: Use Table 9.2, p. 254, in your textbook for assistance.)

   a. Suppose it is a clear, calm, winter night. The stars are shining, the air is dry, and snow covers the ground. Providing the weather does not change, your forecast for tomorrow morning will be:

   b. Suppose the surface wind is blowing from the southeast. It is a cold, winter day with the air temperature in the 20s (°F). You notice high clouds moving in from the west and a halo surrounds the sun. From these observations your forecast for the next 24-hours will be:

   c. Suppose it is a warm summer day in the early afternoon. The sky is dotted with cumulus clouds that have flat bases and tops at just about the same level. Your weather forecast for later this afternoon will be:

Additional Readings

. . . .

"Forecasting into Chaos" by Richard Monastersky, *Weatherwise*, Vol. 43, No. 4 (August 1990), p. 202.

"Warm Snowstorms: A Forecaster's Dilemma" by Stanley David Gedzelman and Elaine Lewis, *Weatherwise*, Vol. 43, No. 5 (October 1990), p. 265.

"The Hi-Tech World of T.V. Weathercasting" by Tom Kierein, *Weatherwise*, Vol. 41, No. 2 (June 1988), p. 150.

"Jerome Namias—Pioneering the Science of Forecasting" by Janet Else Basu, *Weatherwise*, Vol. 37, No. 4 (August 1989), p. 190.

"Dreaming of a White Christmas" by David R. Cook, *Weatherwise*, Vol. 39, No. 6, (December 1986), p. 308.

"Watching the Vapor Channel" by Jack Williams, *Weatherwise*, Vol. 46, No. 4 (August/September 1993), p. 26.

"Forecasts That Fly" by Jack Williams, *Weatherwise*, Vol. 47, No. 6 (December 1994/January 1995), p. 26.

"Feline Forecasters" by Elinor DeWire, *Weatherwise*, Vol. 45, No. 3 (June/July 1992), p. 17.

"Hurd Willett: Forecaster Extraordinaire" by Clifford H. Nielsen, *Weatherwise*, Vol. 46, No. 4 (August/September 1993), p. 38.

"Show and Tell" by Robert Hanson, *Weatherwise*, Vol. 46, No. 5 (October/November 1993), p. 12.

"The Wisdom of Proverbs" by Barbara Houghton, *Weatherwise*, Vol. 49, No. 2 (April/May 1996), p. 26.

"Mapping the Storm" by Mark Monmonier, *Weatherwise*, Vol. 52, No. 3 (May/June 1999), p. 26.

"Sentinels in the Sky" by Jeff Rosenfeld, *Weatherwise*, Vol. 53, No. 1 (January/February 2000), p. 24.

"Model Behavior" by Robert Henson, *Weatherwise*, Vol. 53, No. 3 (May/June 2000), p. 22.

"Forecasting for the Frigid Desert of Antarctica" by Daniel P. Mullen, *Weatherwise*, Vol. 40, No. 6 (December 1987), p. 304.

"The Big Picture," Marshall Shepherd, et al., *Weatherwise*, Vol. 56, No. 1 (January/February 2003), p. 24.

"Weathering History," Carly Kite, *Weatherwise*, Vol. 56, No. 3 (May/June 2003), p. 14.

"Weather When You Really Want It," Dan Stillman, *Weatherwise*, Vol. 56, No. 3 (May/June 2003), p. 22.

"Himalayan Thunder," Yolanda Rosoff, *Weatherwise*, Vol. 56, No. 4 (July/August 2003), p. 12.

"Someone to Blame," Roberta Klein, *Weatherwise*, Vol. 56, No. 5 (September/October 2003), p. 26.

"Mesonet to the Rescue," Angie McNeill, *Weatherwise*, Vol. 57, No. 3 (May/June 2004), p. 34.

"How's the Weather Out There?" by Tim Brookes, *National Geographic*, Vol. 207, No. 6 (June 2005), p. 90.

"Partners in Weather," by Andrew Freedman, *Weatherwise*, Vol. 58, No. 1 (January/February 2005), p. 38.

"Warning: Weather Ahead," by Nick Dalto, *Weatherwise*, Vol. 59, No. 3 (May/June 2006), p. 22.

"Seal of Approval," by Andrew Freedman, *Weatherwise*, Vol. 59, No. 1 (January/February 2006), p. 28.

"Making Records," by Randy Cerveny, *Weatherwise*, Vol. 59, No. 1 (January/February 2006), p. 48.

Answers

Matching

1. h	5. j	9. f	13. g
2. e	6. i	10. b	
3. a	7. d	11. c	
4. k	8. l	12. m	

Fill in the Blank

1. numerical weather prediction
2. climatological
3. warning
4. persistence

Multiple Choice

1. b	4. a	7. e	10. c
2. c	5. e	8. a	11. b
3. d	6. c	9. b	

True–False

1. T	3. T	5. T	7. T
2. F	4. F	6. T	8. T

Additional Questions

1. a. Persistence forecast: "cold and snow."
    b. climatological forecast: "overcast with fog and drizzle."
    c. steady state/trend method: "snow ending, turning colder with clearing weather."

2. a. "Clear and very cold"
    b. "possibility of snow within 12–24 hours; possibly changing to sleet or rain, and windy "
    c. "partly cloudy and warm this afternoon with *no* precipitation, clearing this evening"

Thunderstorms and Tornadoes

10

Chapter Ten examines thunderstorms and tornadoes and the atmospheric conditions that produce them. The beginning of the chapter looks at ordinary (air-mass) thunderstorms and severe thunderstorms. Here we learn that while an ordinary thunderstorm may go through its stages of birth to decay in less than an hour, the supercell thunderstorm, with its updrafts and downdrafts nearly in balance, may exist for hours on end. The chapter also looks at such phenomena as the gust front, the microburst, the squall line, and the mesoscale convective complex. The distribution of thunderstorms as well as lightning and thunder are examined next. This is followed by a section that describes the formation of supercell and nonsupercell tornadoes. The chapter concludes with a discussion of waterspouts.

Some important concepts and facts of this chapter:

1. Ordinary (air-mass) thunderstorms generally form in warm, humid weather, are usually short-lived and rarely produce strong winds and large hail.

2. Severe thunderstorms form in a conditionally unstable atmosphere where the wind speed increases rapidly with height, often in the vicinity of a jet stream.

3. Severe thunderstorms are capable of producing large hail (having at least $3/4$ inch diameter), strong surface winds (with gusts of 50 knots or more), flash floods and tornadoes.

4. Strong downdrafts, called microbursts, have been responsible for several airline crashes because microbursts produce rapid changes in wind speed and wind direction—wind shear.

5. While Florida annually experiences the most thunderstorms, the greatest frequency of hailstorms is over the western Great Plains of the United States.

6. Lightning is a visible electrical discharge—a giant spark—that occurs in mature thunderstorms. It may heat the air through which it travels to 30,000°C.

7. Thunder is the sound that results from the rapidly expanding heated air along the channel of the lightning stroke. Contrary to what a once-popular song claimed, remember: Thunder only happens when it's *lightning*.

8. A tornado is a rapidly rotating column of air around an area of intense low pressure with a circulation that reaches the ground.

9. The visible funnel cloud of a tornado is composed of water (cloud) droplets. The funnel cloud is classified as a tornado when the funnel's circulation reaches the ground.

10. A funnel cloud often descends toward the surface as air rushes into its low pressure core, expands, cools, and condenses at successively lower levels beneath the parent thunderstorm.

11. The majority of tornadoes have winds of less than 125 knots, diameters of less than 600 meters, and occur most frequently in the afternoon.

12. When a tornado approaches, don't bother opening windows—seek shelter immediately!

13. Of all the tornadoes experienced in the United States annually (the average is over 1000), it is the relatively few violent tornadoes that account for the majority of tornado-related deaths.

14. Supercell tornadoes form with supercell thunderstorms—thunderstorms that have a single violently rotating updraft. Nonsupercell tornadoes do not form with supercells. Examples of nonsupercell tornadoes inlude the gustnado and the landspout.

15. Doppler radar has helped scientists gain a great deal of information about tornado-generating thunderstorms.

16. Waterspouts (that form over water) are usually smaller than tornadoes, have weaker winds than tornadoes, and tend to form with developing cumulus clouds.

Self Tests

Match the Following

_____ 1. An enormous thunderstorm that can maintain itself for many hours and produce violent weather

_____ 2. The initial stage of an ordinary (air-mass) thunderstorm

_____ 3. The leading edge of a thunderstorm's cold downdraft

_____ 4. A relatively small downburst, less than 4 kilometers wide

_____ 5. Thunderstorms that develop in a line, one next to the other, each in a different stage of development

_____ 6. This is caused by an aircraft flying faster than the speed of sound

_____ 7. Individual thunderstorms that have grown into a large, long-lasting weather system

_____ 8. The rising, spinning column of air inside a severe thunderstorm

_____ 9. Downdrafts throughout an ordinary (air-mass) thunderstorm are most likely to occur in this stage

_____ 10. Corona discharge and bluish halo that may appear above pointed objects

_____ 11. An ordinary (air-mass) thunderstorm is most intense during this stage

_____ 12. An elongated ominous-looking cloud that often forms just behind a gust front

_____ 13. Rather weak, short-lived tornadoes that often form with building cumulus congestus clouds

a. dissipating

b. microburst

c. mesocyclone

d. supercell storm

e. St. Elmo's fire

f. landspout

g. multicell thunderstorms

h. roll cloud

i. cumulus

j. gust front

k. Mesoscale Convective Complex

l. sonic boom

m. mature

Additional Matching

_____ 1. A tornado cloud whose circulation has not reached the ground

_____ 2. Measures speed at which precipitation is moving horizontally toward or away from you

_____ 3. Rapidly rotating small whirls that sometimes occur with a large tornado

_____ 4. A line of thunderstorms that form along or out ahead of an advancing cold front

_____ 5. Tornadoes that form along a gust front

_____ 6. A tornado-like feature that forms over water

_____ 7. This marks the boundary where warm, moist air encounters warm, dry air

_____ 8. An area of rotating clouds that extends beneath a severe thunderstorm and from which a funnel cloud may appear

_____ 9. The main cause of wind shear associated with several major airline crashes

_____ 10. This often results when intense thunderstorms stall or move very slowly

_____ 11. Extremely strong, damaging, straight-line winds associated with a cluster of severe thunderstorms

_____ 12. Classifies tornadoes according to wind speed and damage

a. flash floods

b. suction vortices

c. Fujita scale

d. wall cloud

e. funnel cloud

f. derecho

g. squall line

h. microbursts

i. Doppler radar

j. waterspout

k. dryline

l. gustnadoes

Fill in the Blank

1. When a tornado is spotted, the National Weather Service issues a _____ _____.

2. A discharge of electricity from or within a cumulonimbus cloud is called _____.

3. The sound produced by rapidly expanding air along the channel of a lightning stroke is called _____.

4. Scattered, isolated, summer thunderstorms that are not severe are called _____ thunderstorms.

5. In cloud-to-ground lightning, the stepped leader travels _____ and the return stroke travels _____.

6. A lightning stroke is seen and 5 seconds later thunder is heard. This means that the lightning stroke is about how many miles away? _____

7. When viewing a severe thunderstorm from the southeast, the most likely place for a tornado to develop is in what section of the thunderstorm? _____

8. Ordinary (air-mass) thunderstorms are most likely to form during what time of the day? _____

9. Lightning associated with thunderstorms that are too far away for the thunder to be heard is referred to as _____ lightning.

10. When it appears that tornadoes are likely to form in a particular area, the National Weather Service issues a _____ _____.

11. On the Fujita scale, the strongest tornado to date was classified as F_____.

Multiple Choice

1. A funnel cloud is composed *mainly* of:

    a. ice particles
    b. hail
    c. cloud droplets
    d. raindrops, dust and dirt from the ground

2. When caught in a thunderstorm in an open field, the best thing to do is to:

    a. run for cover under a tree
    b. stand on your hands for as long as you can
    c. immediately lie down flat on the ground
    d. crouch down as low as possible
    e. stand upright with as little surface area on the ground as possible

3. When lightning illuminates the cloud in which it occurs but its flash can not be seen, the lightning is called:

    a. bead lightning
    b. sheet lightning
    c. ribbon lightning
    d. ball lightning
    e. forked lightning

4.   The majority of tornadoes tend to move from:

    a.   north to south
    b.   northwest to southeast
    c.   south to north
    d.   southeast to northwest
    e.   southwest to northeast

5.   The downdraft in an ordinary thunderstorm is created *mainly* by:

    a.   evaporating raindrops that make the air cold and heavy
    b.   the upper-level wind that dips downward into the thunderstorm
    c.   the release of latent heat as ice particles freeze lightning
    d.   discharges advancing toward the surface
    e.   the melting of snow in the anvil

6.   Tornadoes most frequently form in the:

    a.   middle of the night
    b.   early morning just after sunrise
    c.   early evening just after sunset
    d.   afternoon
    e.   late morning just before noon

7.   Damage is usually most severe during a tornado's:

    a.   dust-whirl stage
    b.   organizing stage
    c.   mature stage
    d.   shrinking stage
    e.   decay stage

8.   You would most likely expect to see St. Elmo's fire:

    a.   over a dry, grassy field
    b.   at the top of a tall, dead tree
    c.   over a plowed, moist field
    d.   over a thick, moist swamp
    e.   near the base of a tree

9.   The funnel cloud characteristic of a tornado is primarily formed by:

    a.   condensation of water vapor that is drawn into the low pressure core of the tornado
    b.   dust and dirt picked up from the surface
    c.   clouds being funneled by downward air currents coming out of a cumulonimbus cloud
    d.   water drawn up from below, into the cloud
    e.   smoke from intense lightning activity within the tornado

10. In the United States, the greatest annual frequency of hail occurs in:

    a. Florida
    b. the Mississippi Valley
    c. the western Great Plains
    d. the Pacific Northwest
    e. Texas

11. In the United States, the greatest annual number of thunderstorms occur in:

    a. Florida
    b. the Mississippi Valley
    c. the Central Plains
    d. the Pacific Northwest
    e. Texas

12. The "Tornado Belt" or "Tornado Alley" of the United States is located:

    a. in Florida
    b. in the Middle Atlantic states
    c. in the Ohio Valley
    d. in the Central Plains
    e. along the Gulf Coast

13. For a thunderstorm to spawn a tornado, the updraft in the cloud must:

    a. be stronger than 100 knots
    b. be saturated
    c. be larger than about 20 kilometers
    d. exist all the way to the top of the cloud
    e. rotate

14. The majority of waterspouts:

    a. draw water up into their core
    b. have rotating winds of less than 45 knots
    c. form beneath severe supercell thunderstorms
    d. form with severe thunderstorms at night
    e. actually form over land

15. Landspouts tend to be similar to:

    a. waterspouts
    b. suction vortices
    c. derechoes
    d. supercell tornadoes

True–False

_____ 1. In a severe thunderstorm, hail may actually fall from the base of the anvil.

_____ 2. All tornadoes make a distinctive roar.

_____ 3. Lightning may occur from one cloud to another.

_____ 4. It is suggested that one *not* open windows as a tornado approaches.

_____ 5. A typical diameter of a tornado would be about one mile.

_____ 6. In the United States, tornadoes are most frequent during the summer and least frequent during the fall.

_____ 7. Different tornadoes spawned by the same thunderstorm are said to occur in families.

_____ 8. All tornadoes rotate counterclockwise.

_____ 9. The air behind the leading edge of a gust front is normally warmer than the air ahead of it.

_____ 10. Thunder only happens when it's raining.

_____ 11. All thunderstorms require rising air.

_____ 12. Only Canada experiences more tornadoes than the United States.

_____ 13. The winds in a typical tornado are usually less than 125 knots.

_____ 14. On a Doppler radar screen a tornado vortex signatue (TVS) appears as a small region of rapidly changing wind direction.

_____ 15. The brightest flash of light during lightning is normally caused by the return stroke.

_____ 16. The top part of a thunderstorm usually has a positive charge.

_____ 17. The "fair-weather" waterspout is normally smaller and less intense than the average tornado.

_____ 18. Nighttime thunderstorms over the Central Plains of the United States appear to be related to a low-level southerly jet stream.

_____ 19. Lightning can momentarily heat the air to 30,000°C, a temperature much hotter than the surface of the sun.

_____ 20. Shelf clouds can form along gust fronts.

_____ 21. A mesohigh is a relatively small area of high pressure created by the cold, heavy air of a thunderstorm's downdraft.

_____ 22. Doppler radar in conjunction with algorithms help forecasters determine which thunderstorms are most likely to produce severe weather.

_____ 23. Gustnadoes and landspouts are examples of nonsupercell tornadoes.

Additional Questions

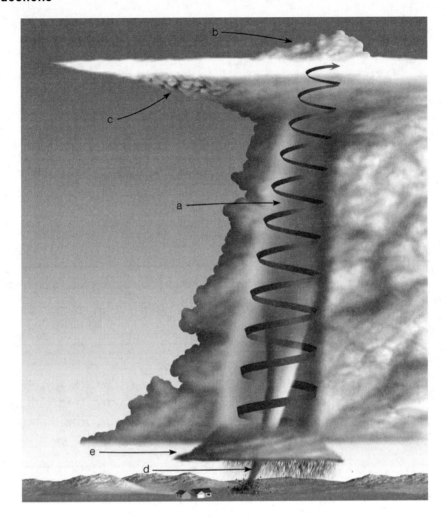

FIGURE 1

1. Figure 1 represents a side view of a severe thunderstorm and each letter represents a feature of this storm. In the space below, place the name of the feature next to each letter.

    a.

    b.

    c.

    d.

    e.

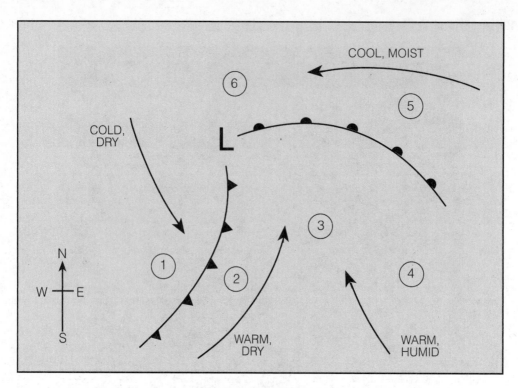

FIGURE 2

2. Figure 2 represents a springtime surface weather map with a warm front and a cold front associated with an open-wave cyclone. Answer the following questions that pertain to this diagram.

   a. The most likely place for a squall line to develop would be near number _____.

   b. The most likely place for severe thunderstorms to form would be near number _____.

   c. A boxed-off area representing a tornado watch would most likely be placed near number _____.

   d. Number 3 represents a boundary called a _____ _____.

   e. If tornadoes should form near number 3, they would most likely move in a direction toward the _____.

   f. Considering the movement of the tornado in (e), the strongest winds should be on which side of the tornado?

      1. southwest side

      2. northwest side

      3. southeast side

      4. northeast side

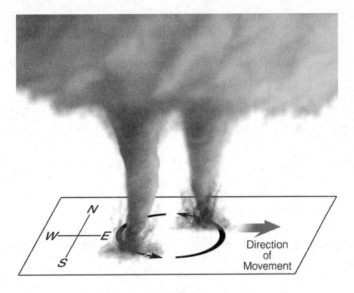

FIGURE 3

3. Suppose the multi-vortex tornado in Figure 3 is moving toward the east at 40 miles per hour. If the counterclockwise rotational wind speed of the tornado is 130 miles per hour and the counterclockwise rotational wind of the suction vortices are 100 miles per hour, then:

    a. What is the maximum wind speed associated with this multi-vortex tornado? _____ (Hint: Maximum wind speed of tornado is the sum of tornado's rotational speed, plus suction vortice's rotational speed, plus forward speed of tornado.)

    b. Would the maximum winds occur on the tornado's north, south, east, or west side? _____

    c. In your textbook, the Fujita scale (Table 10.2, p. 288) ranks this tornado as an F _____. It would fall into what category? _____ The expected damage would be _____.

    d. Would you expect the occurrence of a tornado of this magnitude to be common or rare? _____

Additional Readings

"That's the Way the Building Crumbles" by Hugh Snyder, *Weatherwise*, Vol. 44, No. 3 (June 1991), p. 28.

"The Hesston Tornado: Monster on the Prairie" by Jeff Herzer, *Weatherwise*, Vol. 44, No. 1 (February 1991), p. 23.

"The Tri-State Tornado, March 18, 1925" by Jan Brodt, *Weatherwise*, Vol. 39, No. 2 (April 1986), p. 91.

"Tornado!" by Peter Miller, *National Geographic*, Vol. 171, No. 8 (June 1987), p. 690.

"The Electrification of Thunderstorms" by Earle R. Williams, *Scientific American*, Vol. 259, No. 5 (November 1988), p. 88.

"Flash Floods" by Richard Addison Wood, *Weatherwise*, Vol. 42, No. 2 (April 1989), p. 93.

"Dryline Magic" by Tim Marshall, *Weatherwise*, Vol. 45, No. 2 (April/May 1992), p. 25.

"Rolling Thunder" by Steve Horstmeyer, *Weatherwise*, Vol. 46, No. 6 (December 1993/January 1994), p. 24.

"Elegy for Woodward" by Richard Bedard, *Weatherwise*, Vol. 50, No. 2 (April/May 1997), p. 19.

"The Smell of Tornadoes" by Howard G. Altschule and Bernard Vonnegut, *Weatherwise*, Vol. 50, No. 2 (April/May 1997), p. 24.

"Fires in the Sky" by Daniel Pendick, *Earth*, Vol. 5, No. 3 (June 1996), p. 20.

"Land Spout" by David A. J. Seargent, *Weatherwise*, Vol. 47, No. 3 (June/July 1994), p. 37.

"Tornado Troopers" by Daniel Pendick, *Earth*, Vol. 4, No. 5 (October 1995), p. 40.

"Tornadoes" by Robert Davies-Jones, *Scientific American*, Vol. 273, No. 2 (August 1995), p. 48.

"A Midsummer's Nightmare" by Mace Bentley, *Weatherwise*, Vol. 49, No. 4 (August/September 1996), p. 13.

"Lightning: Serial Killer from the Sky" by Kimbra Cutlip, *Weatherwise*, Vol. 51, No. 4 (July/August 1998), p. 20.

"The Lives of Thunderstorms" by Gregory J. Byrne and Susan K. Runco, *Weatherwise*, Vol. 51, No. 4 (July/August 1998), p. 32.

"Mr. Tornado" by Jeff Rosenfeld, *Weatherwise*, Vol. 52, No. 3 (May/June 1999), p. 18.

"Not Just Flash" by Alan Hall, *Weatherwise*, Vol. 52, No. 3 (May/June 1999), p. 34.

"Wild Waterspouts" by Joseph H. Golden, *Weatherwise*, Vol. 52, No. 5 (September/October 1999), p. 14.

"Riders on the Storm" by Kimbra Cutlip, *Weatherwise*, Vol. 50, No. 5 (October/November 1997), p. 24.

"The Truth About Tornadoes" by Roger Yepsen, *Weatherwise*, Vol. 54, No. 3 (May/June 2001), p. 38.

"Mapping Tornadoes' Deadly Paths" by Allen Carroll, *National Geographic*, Vol. 197, No. 5 (May 2000), p. 132.

"A Stroke of Genius," Nick D'Alto, *Weatherwise*, Vol. 55, No. 3 (May/June 2002), p. 22.

"Tornado Oddities," Randy Cerveny and Joseph T. Schaefer, *Weatherwise*, Vol. 55, No. 4 (July/August 2002), p. 20.

"When Lightning Strikes," Pamela Grim, *Discover*, Vol. 23, No. 8, (August 2002), p. 46.

"Twisting Around the World," Robert Henson, *Weatherwise*, Vol. 56, No. 5 (September/October 2003), p. 14.

"Tornado!" Tim Vasquez, *Weatherwise*, Vol. 56, No. 6, (November/December 2003), p. 32.

"The Business of Chasing Storms," Peggy Willenberg, *Weatherwise*, Vol. 57, No. 1 (January/February 2004), p. 32.

"The Oddities of Lightning," Randy Cerveny, *Weatherwise*, Vol. 57, No. 2 (March/April 2004), p. 64.

"The Hard Science, Dumb Luck, and Cowboy Nerve of Chasing Tornadoes," Priit J. Vesilind and Carsten Peter, *National Geographic*, April 2004, p. 2.

"A Bolt Out of the Blue," by Joseph R. Dwyer, *Scientific American*, Vol. 292, No. 5 (May 2005), p. 64.

"Shelter From the Storm," by Nick D'Alto, *Weatherwise*, Vol. 57, No. 6 (November/December 2004), p. 26.

"Putting Lightning on the Map," by Robert Henson, *Weatherwise*, Vol. 57, No. 5 (September/October 2004), p. 28.

"New View of Tornadoes," by Karen E. Lange, *National Geographic*, Vol. 207, No. 6 (June 2005), p. 110.

"Silent Roar," by David Shiga, *Weatherwise*, Vol. 58, No. 4 (July/August 2005), p. 38.

"California Washed Away: The Grand Flood of 1862," by Jan Null and Joelle Hulbert, *Weatherwise*, Vol. 60, No. 1 (January/February 2007), p. 26

"Fine-Tuning Fujita: A New Tornado Scale Takes Effect," by Sean Potter, *Weatherwise*, Vol. 60, No. 2 (March/April 2007), p. 64.

Answers

Matching

1.	d	5.	g	9.	a	13.	f
2.	i	6.	l	10.	e		
3.	j	7.	k	11.	m		
4.	b	8.	c	12.	h		

Additional Matching

1.	e	4.	g	7.	k	10.	a
2.	i	5.	l	8.	d	11.	f
3.	b	6.	j	9.	h	12.	c

Fill in the Blank

1. tornado warning
2. lightning
3. thunder
4. ordinary (air-mass)
5. downward, upward
6. 1 mile
7. southwest section
8. afternoon
9. heat
10. tornado watch
11. F5

Multiple Choice

1.	c	5.	a	9.	a	13.	e
2.	d	6.	d	10.	c	14.	b
3.	b	7.	c	11.	a	15.	a
4.	e	8.	b	12.	d		

True–False

1.	T	7.	T	13.	T	19.	T
2.	F	8.	F	14.	T	20.	T
3.	T	9.	F	15.	T	21.	T
4.	T	10.	F	16.	T	22.	T
5.	F	11.	T	17.	T	23.	T
6.	F	12.	F	18.	T		

Additional Questions

1.
   a. mesocyclone
   b. overshooting top
   c. anvil (with mammatus)
   d. tornado
   e. wall cloud
2.
   a. 3
   b. 3
   c. 3
   d. dryline
   e. northeast
   f. southeast side
3.
   a. 270 miles per hour
   b. south side
   c. F5, violent, incredible damage
   d. rare occurrence, perhaps only one or two a year

Hurricanes

11

Chapter Eleven focuses on tropical cyclones. The chapter opens with a section on tropical weather. This is followed by several sections that detail a hurricane's structure, formation, dissipation, and movement. Here we learn that hurricanes are tropical cyclones that are given different names in different regions of the world. In the Northern Atlantic and eastern North Pacific they are called hurricanes. In the western North Pacific they are typhoons, and in India and Australia they are tropical cyclones. The chapter concludes by considering how hurricanes are named. The chapter then describes some notable hurricanes of the past. The end of the chapter covers the forecasting and modification of hurricanes.

Some important concepts and facts of this chapter:

1. A hurricane is a tropical cyclone, comprised of an organized mass of thunderstorms, with peak winds about a central core (eye) exceeding 64 knots (74 miles per hour).

2. Winds blow counterclockwise around the hurricane's center in the Northern Hemisphere and clockwise about the center in the Southern Hemisphere.

3. Hurricanes form over tropical waters where light winds converge, the humidity is high, and the surface water temperature is typically 26.5°C (80°F) or greater.

4. Hurricanes derive their energy from the warm tropical waters and from the latent heat released as water vapor condenses into clouds.

5. In the tropics, hurricanes tend to move from east to west.

6. Hurricanes tend to dissipate rapidly when they move over cold water or over a large land mass.

7. Hurricanes are different from mid-latitude cyclonic storms in that hurricanes have warm central cores, an eye where the air is sinking, more closely spaced isobars, no distinct fronts, and usually stronger surface winds.

8. Although the strong winds, huge waves, and high seas of a hurricane can inflict a great deal of damage, it is *flooding* that normally causes the most destruction.

Self Tests

Match the Following

_____ 1.   The zone of intense thunderstorms around the center of a hurricane

_____ 2.   A storm of tropical origin that forms over the North Atlantic and Eastern North Pacific Ocean

_____ 3.   This shows wind-flow patterns on a map

_____ 4.   The center of a hurricane

_____ 5.   What a hurricane is called in India and Australia

_____ 6.   Over the western North Pacific Ocean a storm of tropical origin whose high winds and flooding cause a great deal of destruction

_____ 7.   The stage of hurricane development just after the tropical disturbance (tropical wave) stage

_____ 8.   Weak trough of low pressure in the tropics along which hurricanes occasionally form

_____ 9.   This storm has a low-pressure core and weather fronts

_____ 10. The stage of hurricane development just before it becomes a full-blown hurricane

a.   streamlines

b.   cyclone (tropical)

c.   typhoon

d.   mid-latitude cyclone

e.   eye wall

f.   tropical (easterly) wave

g.   tropical depression

h.   tropical storm

i.   eye

j.   hurricane

Fill in the Blank

1.   Normally, most of the destruction caused by a hurricane is due to _____.

2.   An unusual rise in the ocean level along a shore that is due mainly to the winds of a hurricane: _____ _____.

3.   The _____ _____ scale relates hurricane central pressure and hurricane winds to possible damage hurricanes can cause.

4.   When a hurricane poses a threat to an area, a _____ _____ is issued by the National Weather Service several days before the storm arrives.

5.   For a hurricane to intensify (deepen), the outflow of air at the top of the storm must be _____ than the inflow of air near the surface.

Multiple Choice

1. Which below will *not* cause a hurricane to dissipate?

    a. when it moves over colder water
    b. when it moves over land
    c. when, above the storm, upper-level outflow of air exceeds surface inflow of air

2. The primary source of energy for a hurricane is the:

    a. strong surface winds
    b. heat produced by the sinking of air in the eye
    c. rising of warm air and the sinking of cold air associated with weather fronts
    d. release of latent heat of condensation and warm water
    e. meandering jet stream aloft

3. An area of North America that would *most likely* experience thunderstorms, hurricanes, and tornadoes during the course of a year is the:

    a. region around the Gulf of Mexico
    b. Pacific northwest
    c. south-central Canada
    d. region around the Great Lakes
    e. Great Plains

4. As a hurricane's eye passes directly overhead you would *not* expect to observe:

    a. a rise in temperature
    b. high winds
    c. high clouds
    d. a very low sea level pressure

5. An atmospheric condition that is *not* conducive to the formation of hurricanes is:

    a. strong upper-level winds
    b. an area of organized thunderstorms
    c. warm, humid surface air
    d. warm water
    e. a region of converging surface winds

6. Scientists have tried to modify hurricanes by:

    a. igniting huge smoke bombs in the eye of the storm
    b. seeding the hurricane with hair-thin pieces of aluminum
    c. seeding the hurricanes with silver iodide
    d. placing an oil slick over the ocean water and igniting it

7. A tropical (easterly) wave:

    a. moves from west to east
    b. has showers and thunderstorms on its eastern side
    c. has converging winds on its western side

8.   Hurricanes do *not* form:

    a.   along the equator
    b.   along the intertropical convergence zone
    c.   with a tropical wave
    d.   when the trade wind inversion is weak
    e.   when the surface water temperature is quite warm

9.   Small whirling eddies that form in the strong updrafts of a hurricane's eye wall are known as:

    a.   squall lines
    b.   typhoons
    c.   microbursts
    d.   waterspouts
    e.   spin-up vortices (mini-swirls)

True–False

_____   1.   Hurricanes may contain tornadoes.

_____   2.   The vertical structure of a hurricane shows an upper-level outflow of air, and a surface inflow of air.

_____   3.   A strong trade wind inversion can inhibit the formation of intense thunderstorms and hurricanes.

_____   4.   Tropical cyclones that form over the eastern North Pacific Ocean adjacent to the west coast of Mexico are called hurricanes.

_____   5.   Hurricanes only form over water.

_____   6.   In the center of a hurricane the surface air pressure is much higher than the air pressure around the periphery of the eye.

_____   7.   A hurricane moving north over the Pacific Ocean adjacent to the west coast of North America will normally survive as a hurricane for a longer time than one moving north over the Atlantic Ocean adjacent to the east coast of North America.

_____   8.   Hurricanes usually do not form over the South Atlantic Ocean adjacent to South America because of the relatively cold water found there and the unfavorable position of the ITCZ.

_____   9.   Hurricanes in the Northern Hemisphere are similar to middle latitude cyclones in that both generally move from west to east and have weather fronts.

_____   10.   The winds in a tropical system known as a tropical storm are greater than the winds in a hurricane.

_____   11.   In the eye of a hurricane, several kilometers above the surface, the air is sinking.

_____   12.   Another name for a tropical wave is an easterly wave.

_____ 13. Squall lines do not form in the tropics.

_____ 14. In the heat engine theory of hurricane formation, the maximum strength a hurricane can achieve is determined by the difference in temperature between the ocean surface and the top of the storm's clouds.

Additional Questions

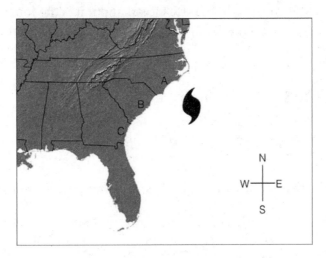

FIGURE 1
Surface Map

1. Suppose the hurricane in Figure 1 has stalled off the southeast coast of North America. Further suppose that the peak winds around the storm are 125 mi/hr and that its central pressure is 934 millibars (27.58 inches). If the hurricane begins to move due west at 15 mi/hr, answer the following questions assuming that the hurricane experiences no change in its intensity.

    a. On which side of the hurricane (north, south, east or west) would the winds be strongest? _____

    b. How strong would these winds be? _____ mi/hr

    c. On which side of the hurricane (north, south, east, or west) would the winds be weakest? _____

    d. How "weak" would these winds be? _____ mi/hr

    e. Along the coastline, would the strongest storm surge most likely be observed at point A, B, or C? _____

    f. Will the hurricane probably make landfall at point A, B, or C? _____

    g. As the hurricane approaches, from which direction will the wind most likely be blowing at point B? _____

    h. With the maximum winds you calculated in (b), and with the storm's central pressure of 934 millibars, on the Saffir-Simpson scale (Table 11.2, p. 313, in your text) determine the category number of this hurricane. _____

i. You would expect a storm surge of about how many feet with this hurricane? _____

j. Flooding with this storm should occur about how many miles inland? _____

k. As the storm moves onshore, will the lowest sea level pressure probably be observed at point A, B, or C? _____

l. If this storm is over the North Atlantic Ocean and if it is the fourth tropical storm or hurricane named in this region during the 2009 hurricane season, what would be the name of this storm? (Hint: Look at Table 11. 1, p. 311, in your textbook.) _____

Additional Readings

"Toward a General Theory of Hurricanes" by Kerry A. Emanuel, *American Scientist*, Vol. 76, No. 4 (July/August 1988), p. 36.

"The Great Galveston Hurricane" by Patrick Hughes, *Weatherwise*, Vol. 43, No. 4 (August 1990), p. 190.

"The Great Hurricane of 1938" by David M. Ludlum, *Weatherwise*, Vol. 41, No. 4 (August 1988), p. 214.

"The Impossible Hurricane—Could It Happen Again?" by A. James Wagner, *Weatherwise*, Vol. 41, No. 5 (October 1988), p. 279.

"The Chesapeake-Potomac Hurricane of 1933" by Hugh D. Cobb, III, *Weatherwise*, Vol. 44, No. 4 (August/September 1991), p. 24.

"Hurricanes Haunt Our History" by Patrick Hughes, *Weatherwise*, Vol. 40, No. 3 (June 1987), p. 134.

"Andrew Aftermath" by Rick Gore, *National Geographic*, Vol. 183, No. 4 (April 1993), p. 2.

"Cities Built on Sand" by Jeff Rosenfeld, *Weatherwise*, Vol. 49, No. 4 (August/September 1996), p. 20.

"Hurricanes in Disguise" by Robert Henson, *Weatherwise*, Vol. 48, No. 6 (December 1995/January 1996), p. 12.

"Inside the Hurricane Center" by Debi Iacovelli, *Weatherwise*, Vol. 47, No. 3 (June/July 1994), p. 28.

"Storm Surge" by Jeff Rosenfeld, *Weatherwise*, Vol. 50, No. 3 (June/July 1997), p. 18.

"The Intensity Problem" by Robert Henson, *Weatherwise*, Vol. 51, No. 5, (September/October 1998), p. 20.

"The Forgotten Hurricane" by Jeff Rosenfeld, *Weatherwise*, Vol. 51, No. 5, (September/October 1998), p. 34.

"Monstrous Mitch" by Mace Bentley and Steve Horstmeyer, *Weatherwise*, Vol. 52, No. 2 (March/April 1999), p. 14.

"Hurricane Camille," *Weatherwise*, Vol. 52, No. 4 (July/August 1999), p. 28.

"Paradise Lost" by Gregory Durrschmidt, *Weatherwise*, Vol. 52, No. 4 (July/August 1999), p. 32.

"Hunting Hugo" by Jeffrey M. Masters, *Weatherwise*, Vol. 52, No. 5 (September/October 1999), p. 20.

"Hurricane Legacies" by Mace Bentley and Steve Horstmeyer, *Weatherwise*, Vol. 52, No. 5 (September/October 1999), p. 28.

"Eye Spies,"Gregory J. Byrne, *Weatherwise*, Vol. 54, No. 4 (July/August 2001), p. 30.

"Modifying the Monsters," Kelli Miller, *Weatherwise*, Vol. 54, No. 6 (November/December 2001), p. 24.

"Awful Agnes," Patricia Barnes-Svarney, *Weatherwise*, Vol. 55, No. 3 (May/June 2002), p. 39.

"Remembering Andrew," Sofia Santana, *Weatherwise*, Vol. 55, No. 4 (July/August 2002), p. 14.

"Moonstruck Meteorology," Joe Rao, *Weatherwise*, Vol. 55, No. 5 (September/October 2002), p. 22.

"Another Charley Horse for Forecasters," by Lee Grenci, *Weatherwise*, Vol. 57, No. 6 (November/December 2004), p. 50.

"Controlling Hurricanes," by Ross N. Hoffman, *Scientific American*, Vol. 291, No. 4 (October 2004), p. 68.

"Hurricane Warning," by Chris Carroll, *National Geographic*, Vol. 208, No. 7 (August 2005), p. 72.

"Katrina Descends," *Weatherwise*, Vol. 58, No. 6 (November/December 2005), p. 10.

"Queen of Rains: Hurricane Camille," by Jeff Halverson, *Weatherwise*, Vol. 58, No. 6 (November/December 2005), p. 24.

"A Climate Conundrum," by Jeffrey B. Halverson, *Weatherwise*, Vol. 59, No. 2 (March/April 2006), p. 18.

"Blown Away," By John L. Beven II, *Weatherwise*, Vol. 59, No. 4 (July/August 2006), p. 32.

"Into the Eye," by Ed Darack, *Weatherwise*, Vol. 59, No. 5 (September/October 2006), p. 46.

Answers

Matching

1. e	4. i	7. g	10. h
2. j	5. b	8. f	
3. a	6. c	9. d	

Fill in the Blank

1. flooding
2. storm surge
3. Saffir-Simpson
4. hurricane watch
5. greater

Multiple Choice

1. c	4. b	7. b
2. d	5. a	8. a
3. a	6. c	9. e

True–False

1. T	5. T	9. F	13. F
2. T	6. F	10. F	14. T
3. T	7. F	11. T	
4. T	8. T	12. T	

Additional Questions

1.
   a. north
   b. 140 mi/hr
   c. south
   d. 110 mi/hr
   e. point A
   f. point B
   g. from the north or northeast
   h. category 4
   i. between 13 feet and 18 feet
   j. about 6 miles
   k. point B
   l. Danny

Air Pollution

12

Chapter Twelve focuses on the important topic of air pollution. The chapter begins with a historical review of air pollution. The various types and sources of air pollutants, including their environmental effects, are covered next. Here we learn that near the surface, ozone—the main component of Los Angeles-type smog—is a secondary pollutant that forms from chemical reactions involving other pollutants in the presence of sunlight. In the stratosphere, however, ozone provides a protective shield against the sun's harmful ultraviolet rays. Here ozone is being depleted by a number of complex chemical reactions and an ozone hole has actually formed over the continent of Antarctica. After examining the trends in air quality across the United States, the chapter examines the different factors that affect the concentration of air pollution, such as wind, atmospheric stability, mixing depth, and local topography. Air pollution and the urban environment is covered next. The chapter concludes with a discussion on acid deposition and its influence on the environment.

Some important concepts and facts of this chapter:

1. Air pollution is not a new problem. It has plagued humanity for a long time.

2. Primary air pollutants enter the atmosphere directly, while secondary air pollutants form when a chemical reaction takes place between a primary pollutant and some other component of air.

3. The major primary pollutants include particulate matter, carbon monoxide, sulfur dioxide, nitrogen oxides, and volatile organic compounds.

4. The word "smog" (coined in London in the early 1900s) originally meant the combining of smoke and fog. Today the word mainly refers to photochemical smog—pollutants that form in the presence of sunlight.

5. The main ingredient of photochemical smog is the secondary pollutant ozone.

6. Photochemical smog forms mainly when the winds are light and the weather is generally warm and sunny.

7. Stratospheric ozone provides protection from harmful shortwave ultraviolet solar radiation.

8.   Most severe air pollution episodes occur when a large high pressure area stalls over a region, bringing with it light winds, a stable atmosphere, a shallow mixing layer, and a strong temperature inversion.

9.   In mountain valleys, air pollution concentrations tend to be greatest during the colder part of the year.

10.   On the average, cities tend to be warmer and more polluted than rural areas.

11.   Emissions of sulfur and nitrogen oxides can be transformed into acids that fall to the surface as acid precipitation.

Self Tests

. . . .

Match the Following

_____ 1. Either solid particles or liquid droplets that remain suspended in the air

_____ 2. Another name for Los Angeles-type smog

_____ 3. This pollutant reacts with water to form nitric acid

_____ 4. A major pollutant of city air, this gas forms during the incomplete combustion of carbon-containing fuels

_____ 5. Another name for a radiation inversion

_____ 6. At night, this blows into a city from the surrounding rural areas

_____ 7. Near the surface, this gas irritates eyes, damages crops, attacks rubber, and forms in the presence of sunlight

_____ 8. A pollutant that comes primarily from the burning of sulfur-containing fuels

_____ 9. Hydrocarbons fall under this category of pollutants

_____ 10. This inversion forms when the air aloft is slowly sinking

_____ 11. The relatively unstable air that extends from the surface to the base of an inversion

_____ 12. Light winds and poor vertical mixing can produce this condition

a. photochemical smog

b. country breeze

c. subsidence inversion

d. atmospheric stagnation

e. volatile organic compounds

f. particulate matter

g. ozone

h. surface inversion

i. mixing layer

j. nitrogen dioxide

k. sulfur dioxide

l. carbon monoxide

Fill in the Blanks

1. Pollutants that form only when a chemical reaction occurs between a primary pollutant and some other component of air are called _____ pollutants.

2. The gas _____ is the main component of Los Angeles-type smog.

3. The index established to indicate the air quality in a particular region is the

   _____ _____ _____.

4. The higher air temperatures of cities contrasted to the cooler surrounding rural areas is referred to as the _____ _____ _____.

5. _____ smog only forms in the presence of sunlight.

6. _____ _____ are those airborne substances that occur in concentrations high enough to threaten the health of people and animals, or to harm vegetation and structures.

7. The _____ determines how quickly air pollutants mix with the surrounding air and how quickly they move away from their source.

8. Acid particles that adhere to fog droplets produce this: _____ _____.

9. When chlorofluorocarbons (CFCs) are subjected to ultraviolet (UV) radiation from the sun, _____ is released, a gas which rapidly destroys ozone.

Multiple Choice

1. Which of the air pollutants below is not emitted *directly* into the air?

   a. sulfur dioxide
   b. nitrogen dioxide
   c. carbon monoxide
   d. particulate matter
   e. ozone

2. In a valley, you would normally expect pollutants to be most concentrated in the:

   a. early morning
   b. early afternoon
   c. early evening

3. Precipitation is usually considered acidic when its pH value is below:

   a. 9
   b. 7
   c. 5
   d. 3

4.   The maximum concentration of ozone is found in this atmospheric layer:

   a.   stratosphere
   b.   troposphere
   c.   thermosphere
   d.   mesosphere

5.   Which smoke plume is responsible for raising the concentration of pollutants to dangerously high levels?

   a.   looping
   b.   fanning
   c.   fumigation

6.   Which below is *not* one of the ingredients for a major build-up of air pollutants?

   a.   light surface winds
   b.   a strong subsidence inversion
   c.   sinking air aloft
   d.   clear skies with rapid radiational cooling at night
   e.   a deep mixing layer

7.   Acid rain can:

   a.   cause a chemical imbalance in the soil
   b.   disfigure outdoor fountains, sculptures, and statues
   c.   damage plants and water resources
   d.   all of the above

8.   At night, as cold air slowly drains into a valley, the cold air normally will *not*:

   a.   strengthen a pre-existing surface inversion
   b.   carry pollutants downhill from the surrounding hillsides
   c.   disperse the pollutants in the valley

9.   Photochemical smog in the Los Angeles area is usually most prevalent during:

   a.   summer and fall
   b.   fall and winter
   c.   winter and spring
   d.   spring and summer

10.   The so-called "ozone hole" is observed above the:

   a.   equator
   b.   continent of Asia
   c.   continent of Antarctica

11. A major pollutant in city air, this gas is the most plentiful of the primary pollutants.

    a. sulfur dioxide
    b. carbon monoxide
    c. nitrogen dioxide
    d. nitric oxide
    e. volatile organic compounds

12. Normally, air pollution episodes are most severe when:

    a. the country breeze is weak
    b. a storm system is developing west of the area
    c. a cold upper-level low moves over the area
    d. a deep high pressure area stagnates over the area
    e. a warm front passes through the area

13. The greatest mixing depth is usually found:

    a. in the early morning, just before sunrise
    b. toward the end of the morning, around noon
    c. in the afternoon

14. Air pollution concentrations in mountain valleys tend to be greatest:

    a. during the colder months
    b. during the warmer months
    c. equally during all the months of the year

True–False

_____ 1. Subsidence inversions are best developed with high pressure areas because of the sinking air associated with them.

_____ 2. Dry deposition is another name for acid rain.

_____ 3. In the early 1900s, the word "smog" meant the combination of smoke and fog.

_____ 4. Transportation vehicles, such as the automobile, account for nearly 50 percent of the pollution across the United States.

_____ 5. Volatile organic compounds (VOCs) include such pollutants as benzene and formaldehyde.

_____ 6. Particulate matter whose diameters are less than 25 micrometers ($PM_{2.5}$) pose the greatest health risk to humans.

_____ 7. On the average, cities tend to be warmer and more polluted than rural areas.

_____ 8. Early air pollution problems were characterized as "smoke problems."

_____ 9. Particulate matter represents a group of pollutants that are only solid, not liquid.

_____ 10. A strong country breeze would probably be associated with a strong heat island.

_____ 11. Air pollution problems have adversely affected humans for only about the last two centuries.

_____ 12. Subsidence inversions normally last longer than radiation inversions.

_____ 13. Until 1990, England was the only country to enact a Clean Air Act.

_____ 14. Motor vehicles represent a mobile source of air pollutants.

_____ 15. On clear, cold winter nights, cities tend to cool more quickly than rural areas and have lower temperatures.

_____ 16. When the base of an inversion lowers, pollutants are able to be dispersed throughout a greater volume of air.

_____ 17. Radiation inversions form at night when the sky is clear and the winds are light.

_____ 18. Emissions of sulfur dioxide and oxides of nitrogen are the pollutants mainly responsible for the production of acid rain.

_____ 19. Normally, taller smoke stacks disperse pollutants downwind more easily than shorter smoke stacks.

_____ 20. The vertical extent of the mixing layer is called the mixing depth.

_____ 21. Sulfurous, smoky air is often called London-type smog.

_____ 22. The concentration of tropospheric ozone in polluted air is much greater than the normal concentration of ozone observed in the stratosphere.

_____ 23. An AQI value of 45 for ozone would be considered unhealthful.

Additional Questions

1. a. In Table 12.2, p. 339, of your textbook (the AQI index), how would the air be described if it had an AQI value of 350 for ozone?

   _____

   b. What would be the general health effects, and what precautions should a person take, with an AQI value of 350?

   c. Should a healthy person go jogging outdoors with an AQI value of 350? _____

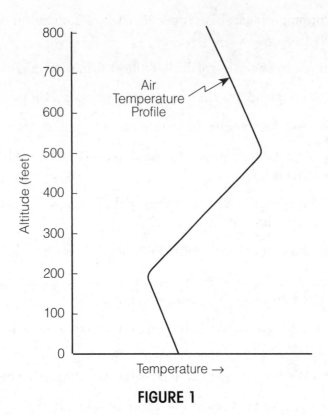

FIGURE 1

2. Figure 1 represents a vertical profile of air temperature from the surface up to 800 feet during the early afternoon. Answer the following questions that pertain to this illustration.

   a. The mixing layer lies below an elevation of _____ feet.

   b. The mixing depth is approximately _____ feet.

   c. The base of the inversion is observed at an elevation of approximately _____ feet, while the top of the inversion is located at about _____ feet.

   d. The greatest concentration in pollutants would be observed from the surface up to an elevation of about _____ feet.

   e. Suppose that by late afternoon the surface temperature increases substantially. Would you expect the mixing depth to increase or decrease? _____ Would the pollutants within the mixing layer become more concentrated or less concentrated?

   _____

Additional Readings

"Smoke Signals" by Alfred Blackadar, *Weatherwise*, Vol. 41, No. 2 (June 1988), p. 159.

"The Challenge of Acid Rain" by Volker A. Mohnen, *Scientific American*, Vol. 259, No. 2 (August 1988), p. 30.

"The Mystery of Arctic Haze" by John Carey, *Weatherwise*, Vol. 41, No. 2 (April 1988), p. 97.

Air Pollution Meteorology by J. R. Eagleman; Trimedia Publishing Company, Lenexa, Kansas, 1991.

"Lost Horizons" by Stephen F. Carfidi, *Weatherwise*, Vol. 46, No. 3 (June/July 1993), p. 12.

Fundamentals of Air Pollution by Richard W. Boubel, D. L. Fox, D. B. Turner, and A. C. Stern; Academic Press, New York, 1994.

"Clearing the Air in Los Angeles" by James M. Lents and William J. Kelly, *Scientific American*, Vol. 269, No. 4 (October 1993), p. 32

"Atmospheric Dust and Acid Rain" by Lars O. Hedin and Gene E. Likens, *Scientific American*, Vol. 275, No. 6 (December 1996), p. 88.

"Wearing Thin" by Scott Fields and Ruth Flanagan, *Earth*, Vol. 3, No. 2 (March 1994), p. 20.

"The Global Transport of Dust: An Intercontinental River of Dust, Microorganisms and Toxic Chemicals Flows Through the Earth's Atmosphere" by Dale W. Griffin, Christina A. Kellogg, Virginia H. Garrison, and Eugene A. Shinn, *American Scientist*, Vol. 90, No. 3 (May 2002), p. 228.

"The Ultimate Clean Fuel: A Start-Up Contemplates Nonpolluting Cars Powered by an Ingredient of Soap" by Julie Wakefield, *Scientific American*, Vol. 286, No. 5 (May 2002), p. 36.

"The Great London Smog," by David Laskin, *Weatherwise*, Vol. 59, No. 6 (November/December 2006), p. 42.

Answers

Matching

1.	f	4.	l	7.	g	10.	c
2.	a	5.	h	8.	k	11.	i
3.	j	6.	b	9.	e	12.	d

Fill in the Blanks

1. secondary
2. ozone
3. air quality index
4. urban heat island
5. photochemical

6. air pollutants
7. wind
8. acid fog
9. chlorine

Multiple Choice

1.	e	5.	c	9.	a	13.	c
2.	a	6.	e	10.	c	14.	a
3.	c	7.	d	11.	b		
4.	a	8.	c	12.	d		

True–False

1.	T	7.	T	13.	F	19.	T
2.	F	8.	T	14.	T	20.	T
3.	T	9.	F	15.	F	21.	T
4.	T	10.	T	16.	F	22.	F
5.	T	11.	F	17.	T	23.	F
6.	T	12.	T	18.	T		

Additional Questions

1. a. hazardous
   b. Premature onset of certain diseases in addition to significant aggravation of symptoms and decreased exercise tolerance in healthy persons; elderly and persons with existing diseases should stay indoors and avoid physical exertion. General population should avoid outdoor activity.
   d. no!

2. a. 200 feet
   b. 200 feet
   c. base 200 feet, top 500 feet
   d. 200 feet
   e. mixing depth would increase; pollutants would become less concentrated

Global Climate

13

Chapter Thirteen examines the large scale aspects of climate. The chapter begins by looking at the global patterns of temperature and precipitation. This is followed by several sections that detail the world's climatic regions. Here we learn that the Köppen scheme of classifying climate employs five major climatic types, each being designated by a capital letter. Tropical moist climates are A, while dry climates are B. Mid-latitude moist climates with mild winters are designated as C, and those with severe winters as D. Polar climates are given the letter E. Each group contains subregions that describe specific regional characteristics, such as seasonal changes in temperature and precipitation. The chapter concludes by examining highland (H) climates—those found in mountainous areas, where rapid changes in elevation bring about sharp changes in climatic zones over a relatively short distance.

Some important concepts and facts of this chapter:

1. Climate is the accumulation of daily and seasonal weather events over a long period of time, as well as the extremes of weather within a specified area.

2. The climatic controls are the factors that govern the climate of any given region. (The seven major climate controls are found on p. 354 in your textbook.)

3. The hottest places on Earth tend to occur in the subtropical deserts of the Northern Hemisphere, where clear skies and sinking air coupled with low humidity and a high summer sun beating down upon a relatively barren landscape, produce extreme heat.

4. The coldest places on Earth tend to occur in the interior of high-latitude landmasses. The Antarctic holds the record for extreme cold.

5. The rainiest places in the world are located on the windward side of mountains where warm, humid air rises upslope.

6. In the middle of a mid-latitude continent, summers are usually wetter than winters.

7. Tropical moist climates are found in low latitudes where abundant rainfall exists, the noon sun is always high, day and night are of nearly equal length, every month is warm, and no real winter season exists.

8. Dry climates prevail where potential evaporation exceeds precipitation. Subtropical deserts, like the Sahara, are mainly the result of sinking air associated with a subtropical high. Other

deserts form on the leeward side of mountains due to the rainshadow effect. Some deserts form in response to both of these effects.

9.   Coastal deserts can experience considerable low cloudiness, fog, and even drizzle.

10. Mid-latitude climates are characterized by distinct winter and summer seasons. Lower latitudes tend to have longer, hotter summers and shorter, milder winters than do higher latitudes. Summers become shorter and winters longer and colder as one moves poleward.

11. Polar climates prevail at higher latitudes where winters are severe and there is no real summer season.

Self Tests

Match the Following

	1.	These plants are capable of surviving a prolonged period of dryness	a.	steppe
	2.	The majority of the southeastern section of the United States has this type of climate	b.	tropical wet
	3.	Name given to the small climatic region near the ground	c.	ocean currents
			d.	laterite
	4.	A grass-covered treeless plain that has a semi-arid climate	e.	microclimate
	5.	One would expect to observe savanna grass in this climate	f.	tropical wet-and-dry
	6.	This is considered a climatic control	g.	humid subtropical
	7.	This type of vegetation grows where the average temperature of the warmest month is above freezing but below 10°C (50°F)	h.	permafrost
			i.	xerophytes
	8.	The freezing of ground to great depths	j.	taiga
	9.	One would expect to observe this type of vegetation in a subpolar climate	k.	tundra
	10.	Abundant yearly rainfall and high temperatures of this climatic type combine to produce a dense, broadleaf, evergreen forest		
	11.	A soil that forms when heavy tropical rains leach nutrients from the soil leaving a hard, red soil that is poor for agriculture		

Fill in the Blanks

1. This climate exists where potential evaporation exceeds precipitation: _____.

2. Another name used to describe a dry-summer subtropical climate: _____.

3. The Köppen scheme for classifying climates employs annual and monthly averages of _____ and _____.

4. This climatic type experiences no distinct summer season: _____ _____.

5. List the seven major climatic controls:

    a. _____

    b. _____

    c. _____

    d. _____

    e. _____

    f. _____

    g. _____

6. This climatic type experiences no distinct winter season: _____ _____.

Multiple Choice

1. A climatic type that experiences a greater temperature variation between day and night than between the warmest and coldest months is:

    a. subpolar
    b. arid
    c. tropical wet
    d. humid subtropical
    e. dry-summer subtropical

2. This climate type covers more land mass than any other:

    a. tropical climates, A
    b. dry climates, B
    c. moist subtropical mid-latitude climates, C
    d. moist continental climates, D
    e. polar climates, E

3. One would expect to observe the melting of an upper layer of permafrost in this climatic type:

   a. polar tundra
   b. polar ice cap
   c. humid continental with warm summers
   d. marine
   e. humid subtropical

4. A city that would most likely experience a dry summer is:

   a. Baltimore, Maryland
   b. Chicago, Illinois
   c. San Francisco, California

5. This region of the world experiences the lowest average temperatures:

   a. the Arctic
   b. Northern Siberia
   c. Northern Canada
   d. Alaska
   e. the Antarctic

6. This climatic type experiences cool, wet winters and mild to hot, dry summers:

   a. humid continental
   b. tropical wet-and-dry
   c. humid subtropical
   d. Mediterranean
   e. tropical monsoon

7. The highest temperatures in the world are normally experienced with this climatic type:

   a. tropical moist climates
   b. dry climates
   c. humid subtropical climates
   d. humid continental climates
   e. highland climates

8. Humid continental (D) climates are not observed in:

   a. Canada
   b. Alaska
   c. Europe
   d. the northern plains of the United States
   e. South America

9. Of the climatic types listed below, which would normally experience the greatest annual range in temperature?

   a.   subpolar climate
   b.   humid subtropical climate
   c.   polar ice cap climate
   d.   arid climates
   e.   tropical wet-and-dry climates

10. The *primary* reason for the dry summer subtropical climate in North America is due to the position of the:

    a.   ITCZ
    b.   polar front
    c.   subtropical high

11. The two primary weather features that influence the tropical wet-and-dry climate are the:

    a.   polar front and ITCZ
    b.   subtropical highs and ITCZ
    c.   subtropical highs and polar fronts
    d.   subtropical highs and subpolar lows
    e.   subpolar lows and ITCZ

12. This climatic type has relatively mild winters and hot, humid summers:

    a.   humid continental
    b.   tropical wet-and-dry
    c.   humid subtropical
    d.   semi-arid
    e.   marine

13. Short bunch grass, scattered low bushes, trees, or sagebrush categorizes a region known as:

    a.   tundra
    b.   taiga
    c.   savanna
    d.   steppe

14. This climatic type experiences adequate precipitation, warm summers, and cold winters with snowstorms and blustery winds:

    a.   humid subtropical
    b.   humid continental
    c.   semi-arid
    d.   polar tundra
    e.   polar ice cap

True–False

_____ 1. One would expect to observe a tropical rain forest in a tropical wet-and-dry climate.

_____ 2. Rainshadow deserts are normally observed on the downwind side of a mountain.

_____ 3. The Köppen scheme for classifying climates does not use the letter H.

_____ 4. The rainiest places in the world are usually located on the windward side of mountains.

_____ 5. Afternoon temperatures in a tropical wet climate are normally much higher than summer afternoon temperatures in the middle latitudes.

_____ 6. In California, one could experience all of Köppen's climatic types.

_____ 7. Deserts that experience low clouds and drizzle tend to be found mainly on the western side of continents.

_____ 8. In a tropical wet-and-dry climate, the dry season occurs in summer, or during the high sun period.

_____ 9. The climate of an area about the size of a city would be described as mesoclimate.

_____ 10. Köppen classified a polar ice cap climate as one where the average temperature of the warmest month averages 32°F or below.

_____ 11. According to Köppen, the climate of a city positioned high in the mountains would be classified as highland.

_____ 12. Most of Canada has a climate classified as polar.

_____ 13. Another name for the taiga climate is boreal climate.

_____ 14. The semi-arid climate marks the transition between the arid and humid climatic regions.

_____ 15. According to Köppen, the driest of all climates would be classified as BW.

Additional Questions

FIGURE 1

1. On the map of the Northern Hemisphere (Figure 1), draw and label the approximate boundaries for Köppen's five major climate regions listed below:

    a. Tropical Moist Climates (Group A)

    b. Dry Climates (Group B)

    c. Moist Subtropical Mid-latitude Climates (Group C)

    d. Moist Continental Climates (Group D)

    e. Polar Climates (Group E)

2. Table 1 contains the mean annual precipitation and mean annual temperature records for five cities. Based on Köppen's climate types, determine the climate of each city. First, determine the major climatic type (A, B, C, D, or E), then the subtype. (Hint: Appendix G on p. 449 in your text will be of assistance with the climate subtype.)

Table 1

		Jan	Feb	Mar	April	May	June	July	Aug	Sept	Oct	Nov	Dec	Year
City 1	temp. (°F)	13	17	26	40	50	56	62	60	51	42	28	19	39
City 1	precip. (in.)	0.5	0.5	0.8	1.0	2.3	3.1	2.5	2.3	1.5	0.7	0.7	0.6	16.5
City 2	temp. (°F)	78	80	82	83	83	81	80	80	80	80	79	78	80
City 2	precip. (in.)	9.1	10.4	9.4	13.0	10.5	6.7	6.4	5.5	10.1	7.2	9.0	10.2	107.5
City 3	temp. (°F)	32	35	43	55	64	74	78	77	70	58	44	35	55
City 3	precip. (in.)	2.0	2.0	3.1	3.7	3.7	4.3	3.3	3.0	2.8	2.9	2.6	2.0	35.4
City 4	temp. (°F)	−26	−26	−13	3	21	37	46	46	37	21	0	−15	11
City 4	precip. (in.)	0.3	0.4	0.4	0.7	0.5	1.0	1.8	1.7	1.5	1.2	0.8	0.6	10.9
City 5	temp. (°F)	31	36	41	48	55	62	70	67	60	51	40	33	50
City 5	precip. (in.)	1.0	1.0	0.7	0.5	0.5	0.4	0.2	0.2	0.2	0.6	0.6	0.9	6.8

Climate for City 1   _____

Climate for City 2   _____

Climate for City 3   _____

Climate for City 4   _____

Climate for City 5   _____

Additional Readings

"Temperature Extremes: A New Cold Champ" by David H. Hickcox, *Weatherwise*, Vol. 44, No. 1 (February 1991), p. 48.

"A December to Remember" by Richard Heim, Kenneth Dewey, Mark Anderson and Daniel Leathers, *Weatherwise*, Vol. 43, No. 6 (December 1990), p. 329.

"The Weather Where You Live" by David A. Robinson, and David M. Ludlum, *Weatherwise*, Vol. 42, No. 6 (December 1989), p. 328.

"How Hot Can It Get?" by David H. Hickcox, *Weatherwise*, Vol. 41, No. 2 (June 1988), p. 157.

"How Often Does It Rain Where You Live?" by Nolan J. Doesken and William P. Eckrich, *Weatherwise*, Vol. 40, No. 4 (July/August), 1987), p. 200.

"Statistical Descriptors of Climate" by Nathaniel B. Guttman, *Bulletin of the American Meteorological Society*, Vol. 70, No. 6 (June 1989), p. 602.

"Valley of the Microclimates" by Steve Horstmeyer, *Weatherwise*, Vol. 47, No. 6 (December 1994/January 1995), p. 13.

"Comfort Zones" by Doug Addison, *Weatherwise*, Vol. 47, No. 3 (June/July 1994), p. 14.

"The 20th Century's Top Ten U. S. Weather and Climate Events" *Weatherwise*, Vol. 52, No. 6 (November/December 1999), p. 14.

"Antarctica: A Land of Ice and Wind" by Jack Williams, *Weatherwise*, Vol. 53, No. 1 (January/February 2000), p. 14.

"The Driest Place on Earth," Priit J. Vesilind and Joel Sartore, *National Geographic*, August 2003, p. 46.

"Hawaii: A Diverse Delight," by David Laskin, *Weatherwise*, Vol. 59, No. 2 (March/April 2006), p. 54.

"Egypt and Water: The Lifeline of the Civilization," by Randy and Niccole Ceerveny, *Weatherwise*, Vol. 59, No. 6 (November/December 2006), p. 42.

Answers

Matching

1. i	4. a	7. k	10. b
2. g	5. f	8. h	11. d
3. e	6. c	9. j	

Fill in the Blanks

1. dry climate
2. Mediterranean climate
3. temperature and precipitation
4. polar climates
5. a. intensity of sunshine and its variation with latitude
   b. distribution of land and water
   c. ocean currents
   d. prevailing winds
   e. positions of high- and low-pressure areas
   f. mountain barriers
   g. altitude
6. tropical climates

Multiple Choice

1. c	5. e	9. a	13. d
2. b	6. d	10. c	14. b
3. a	7. b	11. b	
4. c	8. e	12. c	

True–False

1. F	5. F	9. T	13. T
2. T	6. F	10. T	14. T
3. F	7. T	11. T	15. T
4. T	8. F	12. F	

Additional Questions

1. Check your climatic boundaries on your map (Figure 1), with Figure 13.6 and Figure 13.7 in your textbook.

2. Climate for City 1, D-climate (Dfb)
   Climate for City 2, A-climate (Af)
   Climate for City 3, C-climate (Cfa)
   Climate for City 4, E-climate (ET)
   Climate for City 5, B-climate (BSk)

Climate Change

14

Chapter Fourteen explores the subject of climate change. The beginning of the chapter looks at some of the techniques that have been used to infer past climatic conditions. The next section looks at how the earth's climate has changed in the past. Here we see that during the geologic past, the earth's climate has varied greatly. The following sections describe some of the many theories that attempt to explain why the climate has varied and how the climate may change in the future. At this point we learn that the riddle of the earth's changing climate is not completely understood, because a change in one variable in the complex climate system almost always changes another variable. The chapter concludes with a discussion on recent global warming and the effect increasing concentrations of greenhouse gases have on our climate system.

Some important concepts and facts of this chapter:

1. The earth's climate is constantly undergoing change; studies suggest that the global climate throughout much of the geologic past was much warmer than it is today.

2. The shifting of continents along with volcanic activity and mountain building described by the theory of plate tectonics proposes how climate variation can take place over millions of years.

3. The Milankovitch theory—in association with other natural factors—proposes that small variations in the earth's orbit are responsible for variations in the amount of sunlight reaching the earth and, hence, for glacial advances and retreats during the past 2 million years.

4. Volcanic eruptions, rich in sulfur, may be responsible for cooler periods in the geologic past.

5. Fluctuations in solar output (brightness) may account for climatic changes over time scales of decades and centuries.

6. Based on the predictions of numerical climate models, the IPCC committee in 2001 concluded that increasing levels of greenhouse gases are likely to cause the global mean surface air temperature to increase by between 1.4° and 5.8°C during the period 1990 to 2100. The models also predict, however, that for this amount of warming to occur, the atmospheric concentration of *water vapor* must also increase.

7. At this point, it is not totally certain how clouds and oceans will respond to a warmer earth. Studies show, however, that overall, clouds tend to cool the climate system. A world with fewer clouds would be a warmer world.

8.   Studies indicate that the Earth's mean surface air temperature has warmed by as much as 0.6°C (about 1°F) over the past century, with twelve of the warmest years on record occurring since 1990.

9.   Although no one can unequivocally demonstrate that the recent warming trend is due to increasing levels of CO_2 and other greenhouse gases, it is likely that most of the observed warming during the past 50 years is due to increasing levels of greenhouse gases.

10. In certain regions of the world, overgrazing and deforestation are changing the reflectivity of the surface and, in some cases, rendering the land useless.

Self Tests

. . . .

Match the Following

_____ 1.   The study of annual growth rings of trees

_____ 2.   An increase in the desert conditions of a region

_____ 3.   This refers to a time when there were few sunspots

_____ 4.   When an initial process is reinforced by another process it is referred to as this

_____ 5.   The most potent greenhouse gas

_____ 6.   Another name for the Pleistocene epoch

_____ 7.   This greenhouse gas may more than double in concentration over the next 100 years or so

_____ 8.   Another name for the year 1816

_____ 9.   A time, about 11,000 years ago, when North America and northern Europe reverted back to glacial conditions

_____ 10. Any factor that can change the balance between incoming energy from the sun and outgoing energy from the earth

_____ 11. The circulation of ocean water that plays a role in climate change

a.   Maunder minimum

b.   radiative forcing agent

c.   year without a summer

d.   conveyor belt

e.   Ice Age

f.   positive feedback mechanism

g.   carbon dioxide

h.   dendochronology

i.   desertification

j.   water vapor

k.   Younger-Dryas

Fill in the Blanks

1. Volcanoes that have the greatest impact on global climate appear to be those rich in this gas: _____

2. Climate models predict that in order for increasing levels of CO_2 to raise surface air temperatures 3.5°C by the year 2100, this gas must also increase in concentration: _____.

3. Overall, clouds have a net _____ effect on climate.

4. If the earth were in a warming trend, increasing the amount of water vapor in the atmosphere (but not clouds) would most likely produce a _____ feedback mechanism.

5. A theory that suggests that variations in the earth's orbit are responsible for the advance and retreat of ice over periods of 10,000 to 100,000 years is the _____ theory.

Multiple Choice

1. A negative feedback mechanism:

    a.   weakens an initial change in an atmospheric process
    b.   reinforces an initial change in an atmospheric process
    c.   either reinforces or weakens an initial change in an atmospheric process

2. Most climate models predict that a gradual increase in global CO_2 over the next 100 years will most likely bring about:

    a.   a decrease in surface air temperature
    b.   an increase in surface air temperature
    c.   no change in surface air temperature

3. The theory that explains how glacial material can be observed today near sea level at the equator, even though sea level glaciers probably never existed at the equator, is the:

    a.   volcanic dust theory
    b.   Maunder theory
    c.   theory of plate tectonics
    d.   Milankovitch theory
    e.   changing solar output theory

4. During a period when the earth's orbital tilt is at a minimum, which would probably be true?

    a.   There would be less seasonal variation between summer and winter.
    b.   There would be a lesser likelihood of glaciers in high latitudes.
    c.   In polar regions, less snow would probably fall during the winter.

5. The wobble of the earth on its axis refers to:

    a.   obliquity of the earth's axis
    b.   eccentricity of the earth's orbit
    c.   precession of the earth's axis of rotation

6. If the earth were in a cooling trend, which process below would *most likely* act as a *positive* feedback mechanism?

    a.   decreasing the snow cover around the earth
    b.   increasing the amount of cloud cover around the earth
    c.   increasing the water vapor content of the air

7. The eruption of the Philippine volcano Mount Pinatubo during June, 1991, actually:

    a.   lowered global surface air temperatures
    b.   raised global surface air temperatures
    c.   had no effect on global surface air temperatures

8. Scientists feel that the global warming experienced during the past 50 years is probably the result of:

    a.   increasing levels of CO_2 and other greenhouse gases
    b.   fewer snowfalls and hence a lower surface albedo
    c.   light colored particles in the stratosphere
    d.   a decrease in energy emitted from the sun
    e.   increasing volcanic eruptions

9. Climate models predict that the greatest warming due to increasing levels of greenhouse gases will most likely occur in:

    a.   tropical latitudes
    b.   middle latitudes
    c.   polar latitudes

10. Sulfate aerosols in the lower atmosphere do *not*:

    a.   reflect incoming solar radiation
    b.   serve as cloud condensation nuclei
    c.   decrease the reflectivity (albedo) of clouds

11. As the number of sunspots increases, it appears that the sun's energy output:

    a.   decreases
    b.   increases
    c.   does not change at all

True–False

_____   1.   Studies show that during the past 100 years or so global temperatures have increased.

_____   2.   It appears that throughout much of the earth's history, the climate was much cooler than it is today.

_____   3.   Most of the Sahel in North Africa is a sand-covered desert.

_____   4.   The cooling effect of aerosols resulting from sulfur emissions might have offset part of the greenhouse warming in the Northern Hemisphere during the past several decades.

_____   5.   Studies reveal that during warmer interglacial periods, levels of carbon dioxide were higher than during colder glacial periods.

_____   6.   The Milankovitch cycles, in association with other natural factors, explain how glaciers advance and retreat over periods of ten thousand years to one hundred thousand years.

_____   7.   The CLIMAP project found strong evidence that climatic variations during the past hundred thousand years were closely associated with variations in the sun's energy output.

_____   8.   The higher the ratio of oxygen 18 to oxygen 16 in the shells of organisms that lived in the sea during the geologic past, the colder the climate at that time.

_____   9.   Climate models predict that as global temperatures rise, average global precipitation will increase.

_____   10.   The snow-albedo feedback represents a positive feedback mechanism.

_____   11.   Increasing concentrations of greenhouse gases can be considered radiative forcing agents.

_____   12.   It is not exactly certain at this time how clouds and oceans will respond to a warmer earth.

_____   13.   Extensive glaciation is more likely during periods of cooler summers.

_____   14.   A runaway greenhouse effect is an example of a negative feedback mechanism.

_____   15.   Sulfate aerosols in the lower atmosphere reflect incoming sunlight, which tends to lower the earth's surface air temperature during the day.

_____   16.   Most of the recent global warming over the Northern Hemisphere has occurred at night.

_____   17.   When additional water vapor is added to the atmosphere, the water vapor-greenhouse effect feedback represents a positive feedback mechanism.

Additional Questions

1. In the space provided next to each statement, fill in the blank with the appropriate answer of either "colder" or "warmer." (Assume that only the statement is acting on the atmosphere and that other factors do not come into play.)

    a. Increasing quantities of sulfur-rich volcanic particles in the upper atmosphere should make the surface air _____.

    b. An increase in high-level cirrus type clouds encircling the earth should make the surface air _____.

    c. More ice and snow covering the earth's surface should make the surface air _____.

    d. Higher quantities of water vapor in the atmosphere will likely cause surface air temperatures to be _____.

    e. A decrease in the brightness of the sun will likely cause surface air temperatures to be _____.

    f. If the earth's angle of tilt decreases, summer temperatures in middle latitudes are likely to be _____.

    g. Increasing quantities of CO_2 in the atmosphere will likely cause surface air temperatures to be _____.

    h. If sunspots were to disappear from the face of the sun for an extended period of time, surface air temperatures would likely be _____.

    i. An increase in the low-level stratified clouds encircling the earth is likely to make the surface air _____.

Additional Readings

"Climate Modeling" by Stephen H. Schneider, *Scientific American*, Vol. 256, No. 5 (May 1987), p. 72.

"Drought in Africa" by Michael H. Glantz, *Scientific American*, Vol. 256, No. 6 (June 1987), p. 34.

"Global Climatic Change" by Richard A. Houghton and George M. Woodwell, *Scientific American*, Vol. 260, No. 4 (April 1989), p. 36.

"The Changing Climate" by Stephen H. Schneider, *Scientific American*, Vol. 261, No. 3 (September 1989), p. 70.

"The Great Climate Debate" by Robert M. White, *Scientific American*, Vol. 263, No. 1 (July 1990), p. 36.

"A Second Look at the Impacts of Climate Change" by Jesse H. Ausubel, *American Scientist*, Vol. 79, No. 3 (May/June 1991), p. 210.

"Plateau Uplift and Climatic Change" by William F. Ruddiman and John E. Kutzbach, *Scientific American*, Vol. 264, No. 3 (March 1991), p. 66.

"A Trip to the ICE" by Steve Horstmeyer, *Weatherwise*, Vol. 46, No. 3 (June/July 1993), p. 26.

"The Meteorologist Who Started a Revolution" by Patrick Hughes, *Weatherwise*, Vol. 47, No. 2 (April/May 1994), p. 29.

"Children of the Cold" by Patrick Hughes, *Weatherwise*, Vol. 46, No. 6 (December 1993/January 1994), p. 10.

"Expert Opinion on Climate Change" by William D. Nordhaus, *American Scientist*, Vol. 82, No. 1 (January/February 1994), p. 45.

"Chaotic Climate" by Wallace S. Broecker, *Scientific American*, Vol. 273, No. 5 (November 1995), p. 62.

"The Day the Dinosaurs Died" by Randy Cerveny, *Weatherwise*, Vol. 51, No. 4 (July/August 1998), p. 13.

"Why Has Our Weather Gone Wild?" by Joseph D'Agnese, *Discover*, Vol. 21, No. 6 (June 2000), p. 72.

"Exit from Eden" by Robert Kunzie, *Discover*, Vol. 21, No. 1 (January 2000), p. 84.

"Antarctica's Hot Spot" by Mary Roach, *Discover*, Vol. 20, No. 11 (November 1999), p. 102.

"The Greenhouse Extinction" by Peter D. Ward, *Discover*, Vol. 19, No. 8 (August 1998), p. 54.

"Hot, Hotter, Hottest" by Robert Henson, *Weatherwise*, Vol. 52, No. 2 (March/April 1999), p. 34.

"How Will Climate Change Affect Human Health?" by Pim Martens, *American Scientist*, Vol. 87, No. 6 (November 1999), p. 534.

"Rapid Climate Change" by Kendrick Taylor, *American Scientist*, Vol. 87, No. 4 (July/August 1999), p. 320.

"End of an Ice Age, Onset of a Cold Spell" by Boris Weintraub, *National Geographic*, Vol. 197, No. 2 (February 2000), p. xiii.

"From Tree Rings to Ice Cores," Constanza Villalba, *Weatherwise*, Vol. 53, No. 5 (September/October 2000), p. 20.

"What Wiped Out the Dinosaurs?" Edwin Dobb, et al., *Discover*, Vol. 23, No. 6 (June 2002), p. 36.

"The New Ice Age," Brad Lemley, *Discover*, Vol. 23, No. 9 (September 2002), p. 34.

"Watery Grave," Karen Wright and John Clark, *Discover*, Vol. 24, No. 10 (October 2003), p. 52.

"Can We Bury Global Warming?" by Robert H. Socolow, *Scientific American*, Vol. 293, No. 1 (July 2005), p. 49.

"How Did Humans First Alter Global Climate?" by William F. Ruddiman, *Scientific American*, Vol. 292, No. 3 (March 2005), p. 46

"The Heat Is On," by Tim Appenzeler, et al., *National Geographic*, Vol. 206, No. 3 (September 2004), p. 2.

"Is the Weather Getting Worse?" by David Laskin, *Weatherwise*, Vol. 58, No. 2 (March/April 2005), p. 16.

"Rings of Truth," by Nick D'Alto, *Weatherwise*, Vol. 58, No. 5 (September/October 2005), p. 22.

"When Methane Made Climate," by James F. Kasting, *Scientific American*, Vol. 291, No. 1 (July 2004).

"Going, Going, Gone," by Robert J. Leffler, *Weatherwise*, Vol. 58, No. 3 (May/June 2005), p. 36.

"Spring Forward," by Daniel Grossman, *Scientific American*, Vol. 290, No. 1 (January 2004), p. 84.

"Defusing the Global Warming Time Bomb," by James Hansen, *Scientific American*, Vol. 290, No. 3 (March 2004), p. 68.

"Living on Thin Ice," by Gretel Ehrlich, *National Geographic*, Vol. 209, No. 1 (January 2006), p. 78.

"Meltdown in the Alps," by Erla Zwingle, *National Geographic*, Vol. 209, No. 2 (February 2006), p. 96.

"The Climate Beat," by David Laskin, *Weatherwise*, Vol. 59, No. 4 (July/August 2006), p. 20.

"The Rough Guide to Climate Change: Climate Change and El Niño," by Robert Henson, *Weatherwise*, Vol. 60, No. 1 (January/February 2007), p. 32.

Answers

Matching

1.	h	4.	f	7.	g	10.	b
2.	i	5.	j	8.	c	11.	d
3.	a	6.	e	9.	k		

Fill in the Blanks

1. sulfur
2. water vapor
3. cooling

4. positive
5. Milankovitch

Multiple Choice

1.	a	4.	a	7.	a	10.	c
2.	b	5.	c	8.	a	11.	b
3.	c	6.	b	9.	c		

True–False

1.	T	6.	T	11.	T	16.	T
2.	F	7.	F	12.	T	17.	T
3.	F	8.	T	13.	T		
4.	T	9.	T	14.	F		
5.	T	10.	T	15.	T		

Additional Questions

1. a. colder
    b. warmer
    c. colder
    d. warmer
    e. colder

    f. colder
    g. warmer
    h. colder
    i. colder

Light, Color, and Atmospheric Optics

15

Chapter Fifteen focuses on sunlight in our atmosphere and the visual effects it produces. The chapter begins by describing how scattered sunlight can produce an array of atmospheric visuals from white clouds to blue skies to crepuscular rays. The next several sections discuss how the bending of light produces such phenomena as mirages, halos, and sundogs. After this we see that rainbows are the result of light being bent, reflected, and dispersed. The later part of the chapter describes how coronas and cloud iridescence form.

Some important concepts and facts of this chapter:

1. Light that is scattered is sent in all directions—forward, sideways, and backwards.

2. An object that appears white, yet is not hot enough to generate its own light, has all visible wavelengths reflected or scattered from its surface. An object that technically has little or no visible light returning from its surface appears black.

3. Blue skies are the result of air molecules scattering the shorter waves of visible light (blue) more than the longer visible waves (red).

4. The bending of light as it moves through regions (objects) of different density is called refraction.

5. The dispersion of light separates white light into its different component colors.

6. For a rainbow to form, rain must be falling in one part of the sky and the sun must be shining in another.

7. The bending of light as it passes around objects is called diffraction.

Self Tests

Match the Following

_____	1.	You can see this only when the sun is at your back and it is raining in front of you	a.	halo
_____	2.	Phenomenon that makes objects appear higher or lower than they actually are	b.	green flash
			c.	mirage
_____	3.	The time after sunset or before sunrise when the sky is illuminated	d.	cloud iridescence
_____	4.	Another name for diffuse light	e.	rainbow
_____	5.	A ring of light encircling and extending outward from the sun or moon	f.	scattered light
			g.	sun pillar
_____	6.	Beams of light shining downward through breaks in clouds	h.	scintillation
_____	7.	A vertical streak of light extending above (or below) the sun	i.	crepuscular rays
			j.	twilight
_____	8.	A ring of light that appears to rest on the moon or sun	k.	corona
_____	9.	This can be seen near the upper rim of the sun at sunrise or sunset		
_____	10.	The apparent twinkling of a star		
_____	11.	Clouds that exhibit patches of color, often pastel shades of pink and blue		

Additional Matching

Match the following with the process that is mainly responsible for its formation. (Note: some answers will be used more than once.)

_____ 1. crepuscular rays

_____ 2. inferior mirage

_____ 3. halo

_____ 4. blue skies

_____ 5. corona

_____ 6. sun pillar

_____ 7. white clouds

_____ 8. star scintillation

_____ 9. hazy skies

_____ 10. blue moons

_____ 11. wet-looking pavement on a clear, dry, hot day

_____ 12. Fata Morgana

_____ 13. tangent arc

_____ 14. cloud iridescence

_____ 15. sundog

a. scattering

b. reflection

c. refraction

d. diffraction

Fill in the Blanks

1.   If there was no atmosphere surrounding the earth, the *sky* during the day would appear what color? _____

2.   When sunlight strikes an object and the light is sent in all directions, the light is said to be _____.

3.   The bending of light through an object is called _____.

4.   The breaking up of white light by selective refraction is called _____.

5.   The type of mirage that makes objects appear higher than they actually are is called a _____ _____.

6.   For halos and sun pillars to form, _____ _____ must be present in the atmosphere.

Multiple Choice

1.   If you were standing on the moon, the color of the *sky* would be:

   a.   white
   b.   red
   c.   blue
   d.   purple
   e.   black

2.   Which of the following processes must occur in a raindrop to produce a rainbow?

   a.   refraction, reflection, and scattering of sunlight
   b.   reflection, scattering, and dispersion of sunlight
   c.   refraction, reflection, and dispersion of sunlight
   d.   transmission, scattering, and dispersion of sunlight
   e.   refraction, transmission, and scattering of sunlight

3.   As light passes through ice crystals, _____ light is bent the *least* and is, therefore observed on the _____ of halos and sundogs.

   a.   red, outside
   b.   red, inside
   c.   blue, inside
   d.   blue, outside

4.   The bending of light as it passes *around* objects is called:

   a.   scattering
   b.   dispersion
   c.   diffraction
   d.   refraction
   e.   reflection

5. On a foggy night it is usually difficult to see the road when the high beam lights are on because of the _____ of light.

    a. scattering
    b. absorption
    c. transmission
    d. refraction

6. If you looked in the general direction of the sun you would not be able see a:

    a. corona
    b. halo
    c. sundog
    d. rainbow
    e. sun pillar

7. If the setting sun appears red, you may conclude that:

    a. the sun's surface temperature has changed
    b. the next day will be extremely hot
    c. only the longest waves of visible light are striking your eye
    d. the moon will not shine
    e. there is something in your eye

8. Clouds in the tropics tend to move from east to west. Consequently, which rhyme best describes a rainbow seen in the tropics?

    a. Rainbow at the break of day/ means that rain is on the way
    b. Rainbow in morning/ sailors should take warning
    c. Rainbow with a setting sun/ means that sailors can have some fun
    d. Rainbow at the break of dawn/ means, of course, the rain is gone

9. When distant mountains appear blue, it is mainly due to the _____ of light.

    a. scattering
    b. dispersion
    c. refraction
    d. scintillation

True–False

_____ 1. Rainbows form in much the same way as a mirage.

_____ 2. At sunrise in the middle latitudes, a sundog would appear in the west.

_____ 3. The Fata Morgana is actually a type of mirage.

_____ 4. When a star appears near the horizon, its actual position is slightly higher.

_____ 5. The best time of day to see a rainbow is around noon.

_____ 6. If the earth had no atmosphere, the stars would be visible all the time.

_____ 7. Small suspended salt particles, volcanic ash, and small suspended dust particles are all capable of producing red sunrises and sunsets.

_____ 8. Secondary rainbows occur when two internal reflections of light occur in raindrops.

_____ 9. The best time of day to see the green flash is around noon when the sun's rays are most intense.

_____ 10. The Heiligenschein is most easily seen at night.

Additional Questions

1. In the space provided below, place the name of the optical phenomenon that appears in the adjacent figures.

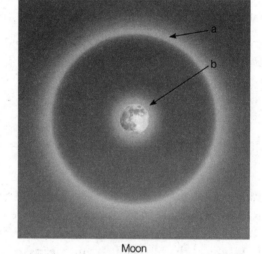

Moon

FIGURE 1

a.

b.

Sun

FIGURE 2

c. _____

d. _____

FIGURE 3

e. _____

FIGURE 4

f. _____

FIGURE 5

g. _____

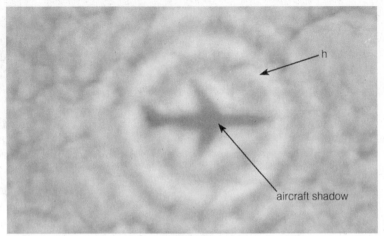

FIGURE 6

h. _____

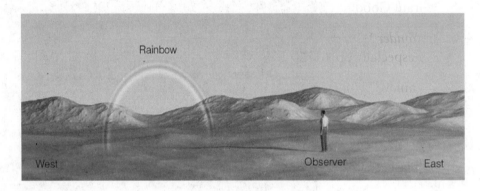

FIGURE 7

2. In Figure 7:

    a. Place the sun in its proper position for the rainbow to appear as shown.

    b. Place the clouds and falling rain in their proper position.

    c. If the person observing the rainbow is located in the Central Plains of the United States, based on the information in the illustration, would the person expect clearing weather or showers? Explain.

Additional Readings

"Colors of the Sky" by Craig F. Bohren and Alistair Fraser, *Physics Teacher*, Vol. 23, No. 5 (May 1985), p. 267.

"The Boulder, Colorado, Concentric Halo Display of 21 July 1986" by Paul J. Neiman, *Bulletin* of *the American Meteorological Society*, Vol. 73, No. 3 (March 1989), p. 258.

Rainbows, Halos, and Glories by Robert Greenler; Cambridge University Press, New York, Cambridge, 1980.

Clouds in a Glass of Beer by Craig F. Bohren; John Wiley and Sons, Inc., New York, 1987. See especially pp. 86–185.

"Sundogs" by T. D. Nicholson, *Natural History*, (February 1991), p. 70.

"Solar Circles" by Grant Goodge, *Weatherwise*, Vol. 45, No. 3 (June/July 1992), p. 8.

What Light through Yonder Window Breaks? by Craig F. Bohren; John Wiley and Sons, Inc., New York, 1991. See especially pp. 61–70 and 165–177.

"Halo Heaven" by Stanley David Gedzelman, *Weatherwise*, Vol. 48, No. 4 (August/September 1995), p. 34.

"Jewels of the Sky" by Russell D. Sampson, *Earth*, Vol. 5, No. 5 (October 1996), p. 54.

"Stalking the Green Flash!" by Mark J. Coco, *Weatherwise*, Vol. 49, No. 6 (December 1996/January 1997), p. 31.

"The Colors of Twilight" by Stephen Corfidi, *Weatherwise*, Vol. 49, No. 3 (June/July 1996), p. 14.

"Ray from Heaven" by Mark Schneider, *Weatherwise*, Vol. 50, No. 6 (December 1997), p. 31.

"When and Why the Sun Turns Green," *National Geographic*, Vol. 198, No. 5 (November 2000), p. xxii.

"Colors of the Sky," Stanley David Gedzelman, *Weatherwise*, Vol. 55, No. 1 (January/February 2002), p. 20.

"Every Dog Has Its Day," Stephen R. Wilk, *Weatherwise*, Vol. 55, No. 6 (November/December 2002), p. 34.

"Capturing Rainbows," Bill Speaks, *Weatherwise*, Vol. 57, No. 2 (March/April 2004), p. 14.

"The Useful Pursuit of Shadows," by Graema L. Stephens, *American Scientist*, Vol. 91, No. 5 (September/October 2003), p. 442.

"Why We See What We Do," by Dale Purves, et al., *American Scientist*, Vol. 90, No. 3 (May/June 2002), p. 236.

"Spectre Rising," by Niccole and Randy Cereveny, *Weatherwise*, Vol. 57, No. 4 (July/August 2004), p. 38.

Answers

Matching

1.	e	4.	f	7.	g	10.	h
2.	c	5.	a	8.	k	11.	d
3.	j	6.	i	9.	b		

Additional Matching

1.	a	5.	d	9.	a	13.	c
2.	c	6.	b	10.	a	14.	d
3.	c	7.	a	11.	c	15.	c
4.	a	8.	c	12.	c		

Fill in the Blank

1.	black	4.	dispersion
2.	scattered	5.	superior mirage
3.	refraction	6.	ice crystals

Multiple Choice

1.	e	4.	c	7.	c
2.	c	5.	a	8.	d
3.	b	6.	d	9.	a

True–False

1.	F	4.	F	7.	T	10.	F
2.	F	5.	F	8.	T		
3.	T	6.	T	9.	F		

Additional Questions

1.
   a. halo
   b. corona
   c. sun pillar
   d. sundog
   e. tangent arc
   f. crepuscular rays
   g. inferior mirage
   h. glory

2.
   a. Sun's position would be to the east of the observer.
   b. The clouds and falling rain would be located directly above the rainbow.
   c. Observer could expect showers because at this location clouds tend to move from west to east.